Assembling Life

Assembling Life

*How Can Life Begin on Earth
and Other Habitable Planets?*

DAVID DEAMER

OXFORD
UNIVERSITY PRESS

OXFORD
UNIVERSITY PRESS

Oxford University Press is a department of the University of Oxford. It furthers
the University's objective of excellence in research, scholarship, and education
by publishing worldwide. Oxford is a registered trade mark of Oxford University
Press in the UK and certain other countries.

Published in the United States of America by Oxford University Press
198 Madison Avenue, New York, NY 10016, United States of America.

Library of Congress Cataloging-in-Publication Data
Names: Deamer, David W., author.
Title: Assembling life : how can life begin on earth and other habitable planets? /
David W. Deamer.
Description: New York, NY : Oxford University Press, [2019] |
Includes bibliographical references and index.
Identifiers: LCCN 2018030319 | ISBN 9780190646387 (hardback : alk. paper) |
ISBN 9780190646394 (updf) | ISBN 9780190646400 (epub)
Subjects: LCSH: Life—Origin. | Evolution (Biology) | MESH: Origin of Life |
Biological Evolution | Exobiology | Biochemical Phenomena | Geological Phenomena
Classification: LCC QH325.D74 2019 | NLM QH 325 | DDC 576.8/3—dc23
LC record available at https://lccn.loc.gov/2018030319

CONTENTS

PREFACE

In 1964, long before the internet gave instant access to the accumulated knowledge of humankind, it was necessary to actually visit a library full of books, thumb through a thick volume of *Chemical Abstracts* to find the desired citation, and then trudge up several flights of stairs to crack open the bound volume in which the desired paper was embedded. The article I wanted was in *Archives of Biochemistry and Biophysics*, and while leafing through the pages, I happened to see the title "Synthesis of Purines Under Possible Primitive Earth Conditions." As a novice biochemist, I knew that purines were components of nucleic acids, but I had never before heard the phrase "primitive earth." What could that mean? I began to read and was fascinated. Someone named John Oro at the University of Houston had discovered that adenine was a pentamer of hydrogen cyanide (HCN) and that this simple reaction could have provided one of the essential nucleobases required for life to begin. My curiosity satisfied, I found the paper I needed, took notes by hand because photocopiers had not yet been invented, and then returned to the lab where I was studying how calcium ions interact with fatty acid monolayers. I would never have imagined that fifteen years later I would be visiting Oro in Houston and writing a paper with him on how lipid membranes could have been involved in life's beginning.

The study of life's origins was, in the 1960s and 1970s, on the fringes of scientific research and may still be yet, although inroads are being made. As I went through post-doctoral studies at University of California (UC), Berkeley and then began my academic career at UC Davis, I did not give it another thought. The topic was definitely not something a young assistant professor who wanted to impress senior colleagues and be awarded a tenured position should get interested in. Instead, my research was focused on biological membranes—of mitochondria, chloroplasts, and sarcoplasmic reticulum—using the new technique of freeze-fracture electron microscopy to study their structure and function.

By 1975 I was a tenured associate professor and decided to celebrate by taking a sabbatical to learn more about the membrane models called liposomes. These models had been developed by the British researcher Alec Bangham, so our little family of four traveled to Cambridge, England, found a home to rent, and every morning I rode a red, double-decker bus to Babraham where Alec had his laboratory at the Animal Physiology Institute. I have much to say about the discoveries we made over the next six months, too much for this preface. What I will recount is a conversation with Alec while we were driving down to London in his Morris Mini. I don't recall the details, but somewhere along the way we stopped for a picnic lunch alongside the road. I was curious about a talk Alec had given at the University of Bristol with the title "Membranes Came First" (Fig. P.1). During our conversation we realized that there was a vast gap in our understanding: What were the first lipids and how did they assemble into membrane compartments required for the first forms of life? I took that question home with me.

Forty years later, the seed planted by that question has sprouted and grown into what might be called a shrub. In other words, I can describe multiple branches related to a few answers to the main question, but there is not yet a central trunk of consensus that could be called a tree. This book describes the growth process of this shrub in a certain way, because during those forty years I also wrote two textbooks, the first about human physiology and the second about cell biology. The textbook style, for better or worse, permeates the style that I have chosen to present information in this book. In other words, the author of a textbook tries to be as clear as possible by providing an introduction to each chapter, a list of questions to be addressed, and then the main discussion followed by a summary with conclusions at the end. My impression is that even experts in origins of life research will benefit from the textbook style. Certainly it helped me keep my thought processes on track as I wrote.

As a final word, I must confess to writing an earlier book about the origin of life. *First Life* was published in 2010 by the University of California Press. Just eight years later, why should I write another? *First Life* was what is called a trade book, written for readers

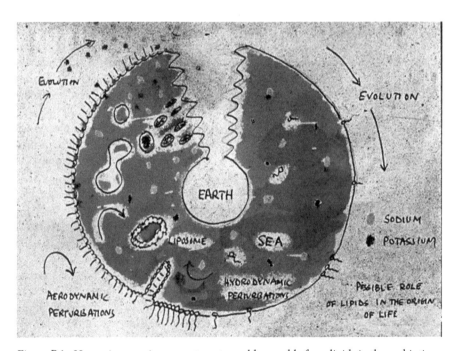

Figure P.1. How microscopic compartments could assemble from lipids in the prebiotic ocean. Credit: Alec Bangham; from a lecture presented in 1971.

who did not have scientific training but were still interested in the question of life's beginning. *Assembling Life* is a heftier monograph written for the scientific community. Why heftier? Well, throughout the book there are citations to the scientific literature. There are even a few equations in Chapter 6! But more important, a lot has happened in the eight years since *First Life* was published, such as the discovery of liquid water on Enceladus and the consensus that Mars once had lakes and rivers. What are the chances that life could begin elsewhere on other planetary bodies in the solar system? That alone is worth writing about.

I also tried to make this book interesting and readable. It is not intended to be a highly detailed compendium reviewing research on the origin of life with a thousand or more citations in a reference section. Instead, the book describes a personal map of largely unexplored territory that encompasses the research I have carried out with colleagues and in collaboration with talented undergraduates, graduate students, and post-doctoral scholars. When I do refer to the research of other scientists, the citations are landmarks that have guided the map-making process. The goal is to construct a coherent narrative in the form of a testable hypothesis that suggests a set of critical experiments. My intention is not to show how life *did* begin on Earth, because we can never know that with certainty. Instead, the goal is to propose how life *can* begin on any habitable planet, with the Earth so far being the only known example.

QUESTIONS THAT GUIDE THE CONTENT AND LOGIC OF CHAPTERS

ACKNOWLEDGMENTS

This book is dedicated to the memory of Alec Bangham and Harold Morowitz. Alec demonstrated the principles of membrane self-assembly with soap bubbles and liposomes, and Harold made it abundantly clear that the origin of life could only be understood in the light of bioenergetics.

I want to thank Dan Branton and Mark Akeson for 25 years of friendship as we pushed nanopore sequencing through to seeing it incorporated into a working device. Results from nanopore analysis of RNA-like polymers are described in Chapter 9.

Vladimir Kompanichenko convinced me to travel to Kamchatka where I first realized how important it was to experience prebiotic-Earth-analogue conditions and watch what happens.

Particular appreciation goes to Bruce Damer, a perfect example of serendipity as we discovered a mutual passion for working in Western Australia and seeing for ourselves the ancient stromatolite fossils, clear evidence that life goes back to at least 3.4 billion years ago. I invited Bruce to co-author Chapter 10. Tara Djokic, whose discovery of geyserite in the Dresser formation led to a revelation about early life, provided an expert critique of Chapter 1, and Miguel Aznar volunteered to read the entire manuscript for clarity.

I also want to thank David Ross for co-authoring Chapter 6. Dave spent his scientific career at SRI using thermodynamic and kinetic theory to guide studies of organic reactions, and he brought a healthy skepticism to the idea that polymerization can be driven by simple cycles of wetting and drying. Our collaboration resulted in the conclusion that it was thermodynamically feasible, an important test that should be a foundation for any hypothesis claiming to explain the origin of life.

National funding agencies like the National Institutes of Health (NIH) dole out several billion dollars every year to support research related to health and disease. The only source of funds for origins of life research is the National Aeronautics and Space Administration (NASA), but the dollar amounts made available amount to a few millions, not billions. Fortunately, private donors help fill the gap. Harry Lonsdale was a retired chemist from Bend, Oregon, and generously supported my research from 2012 until his death in 2015. He once told me that his gift marked his belief that the origin of life is not an impenetrable mystery but will someday be understood in terms of chemical and physical laws.

Then there are my colleagues from whom I learned so much as we worked together and published together: Lou Allamandola, Roy Black, Ernesto DiMauro, Jason Dworkin, Pascale Ehrenfreund, Gail Fleischaker, Christos Georgio, Bob Hazen, Jerry Joyce, Doron Lancet, Pier Luigi Luisi, Marie-Christine Maurel, Pierre-Alain Monnard, John Oro, Andrew Pohorille, Sudha Rajamani, Maikel Rheinstadter, Daniel Segre, Meir Shinitzky, Bernd Simoneit, Martin Van Kranendonk, Art Weber and Dick Zare.

I am also indebted to another group of colleagues. We never happened to collaborate on publications but much of what I wrote about in this book comes from hearing their talks at meetings, reading their papers, visiting their labs, and enjoying their company: John Baross, Jeff Bada, Steve Benner, Chris Chyba, George Cody, Freeman Dyson, Jim Ferris, George Fox, Paul Higgs, Nick Hud, Stu Kauffman, Noam Lahav, Ram Krishnamurthy, Nick Lane, Antonio Lazcano, Lynn Margulis, Stanley Miller, Leslie Orgel, Ralph Pudritz, Lynn Rothschild, Carl Sagan, Everett Shock, Norm Sleep, Günther Wächtershäuser, Sandra Pizzarello, Jack Szostak, Malcolm Walter, Loren Williams, and Kevin Zahnle

Every scientist knows that without graduate and undergraduate students, our labs would be empty and not much would get done. Charles Apel, Gail Barchfeld, Ajoy Chakrabarti, Laura Da Silva, Tara Djokic, Veronica DeGuzman, Matt Groen, Will Hargreaves, Chrissa Karagianis, Augustin Lopez, Ryan Lorig-Roach, Sarah Maurer, Wylie Nichols, Felix Olasagasti, Stefan Paula, Laura Toppozini, and Wenonah Vercoutere. Thank you for sharing years of your lives with me as we toiled through the nine out of ten experiments that failed and rejoiced in the one that worked.

And thank you, Ólöf, Ásta, and Stella, for being my family, for our visits to Iceland and Hawaii that taught me so much about volcanoes, and for putting up with the long hours in the study as I wrote this narrative.

INTRODUCTION

> It was unthinkable not long ago that a biologist or paleontologist would
> be at the same conference as an astrophysicist. Now we have accumulated
> so much data in each of these branches of science as it relates to origins
> that we have learned that no one discipline can answer questions of
> origins alone.
>
> Neil deGrasse Tyson

Conjectures, Hypotheses, and Consensus

In origins of life research, it is important to understand the difference between conjecture and hypothesis. All ideas begin as conjectures, defined as opinions based on insufficient evidence. A conjecture becomes a hypothesis when it is possible to devise experiments or observations that test an idea, or falsify it—using the terminology of Karl Popper. All too often two scientists come up with conjectures that are incompatible in some way and controversy ensues. My advice to the scientists engaged in such a controversy is not to argue back and forth with claims of correctness and plausibility but instead, to accept that the two ideas are best investigated as alternative hypotheses as first defined by John Platt (1964). Testing alternative hypotheses avoids the human tendency to fall in love with and try to protect a precious idea. Instead, the weight of evidence that accrues from the tests gives value to both positive and negative results. At some point the weight of evidence convinces colleagues, just as evidence convinces jurors in a trial, and a consensus is reached that one of the alternative hypotheses has more explanatory power.

This book will describe a number of aspects regarding the origin of life that are currently controversial. For instance, which came first—metabolism or genes? Can life exist in the absence of membranous compartments? More specifically, did life begin in the ocean, in salty seawater, or in freshwater on terrestrial land masses? The last question is particularly important because an answer has ramifications related to the origin of life on other planetary bodies that will be described in Chapter 12.

To make this clear, I will show how alternative conjectures can arise from a factual basis (adapted from Deamer, 2017). The first conjecture follows from the discovery of hydrothermal vents and observations related to their properties:

Facts:

- All life requires liquid water.
- Most of the water on Earth is in the ocean.

- Hydrothermal vents emerging from the ocean floor are sources of chemical energy.
- Populations of chemotrophic microbial life thrive in hydrothermal vents.

Conjecture: Life originated in hydrothermal vents and later adapted to freshwater on volcanic and continental land masses. In the absence of alternatives this idea has been accepted as a reasonable proposal.

Here is an alternative to a marine origin of life.

Facts:

- A small fraction of the Earth's water is distilled from seawater and precipitates as freshwater on volcanic land masses.
- The water accumulates in hydrothermal fields that undergo cycles of evaporation and refilling.
- During evaporation, dilute solutes in the water become concentrated films on mineral surfaces.
- If the solutes can undergo chemical or physical interactions, they will do so in the concentrated films.
- The products will accumulate in the pools when water returns either in the form of precipitation or as fluctuations in water levels related to hot springs or geyser activity.

Conjecture: Life originated in freshwater hydrothermal fields associated with volcanic land masses then later adapted to the salty seawater of the early ocean.

Throughout this book, I will be treating these conjectures as alternatives and will present the weight of evidence related to each hypothesis.

The Early Earth: Facts, Assumptions and Plausibility

I don't recall who told me this, but years ago I was given sound advice about presenting scientific talks, which goes something like this: "First tell 'em what you'll tell 'em, then tell 'em, then tell 'em what you told 'em." Following that advice, here is my personal scenario for the first three chapters of this book, based on idiosyncratic estimates of the plausibility of the various conditions. Not everyone will agree with the guesswork, but we must start somewhere in order to design experimental tests of ideas.

- The first physical and chemical steps toward living systems began after the global temperature had fallen sufficiently for an ocean to condense from water vapor. Therefore it is likely that cellular life emerged in conditions much warmer than today's global temperature of $15°$ C.
- The ocean was already salty, with volcanic land masses emerging into an atmosphere mostly composed of nitrogen and a smaller fraction of carbon dioxide.
- The land masses resembled Hawaii and Iceland today, caused by magma plumes heating the Earth's crust and producing volcanic islands. Evidence for similar ancient volcanoes on Mars include Apollinaris Mons, estimated to have been active ~3.5 billion years ago.
- Hydrothermal chemistry is a plausible way to develop the complexity required for life to begin. Two hydrothermal sites have been proposed as sites for the origin of

life: submarine hydrothermal vents and hydrothermal fields associated with volcanic land masses.

- Hydrothermal vents emerge into salty seawater and have free energy available in the form of reducing power, pH gradients, and thermal gradients.
- Hydrothermal fields associated with volcanism are supplied with fresh water distilled from salty seawater and falling onto land masses as precipitation. Contemporary examples include the hydrothermal fields I have investigated in Kamchatka, Hawaii, Iceland, New Zealand, and Yellowstone National Park. Based on what we see today, it is reasonable to assume that the earliest pools were heated by geothermal energy and ranged in size from puddles with volumes of a few liters to larger ponds fed by geysers, hot springs, and cyclic precipitation.

With those assumptions in place, a set of biological properties can be presented as a list that illustrates the complexity of life and brings us to a fundamental question of biology.

All life today:

- Exists in a steady state away from equilibrium
- Is based on polymers that are synthesized by condensation reactions
- Uses energy from the environment to drive polymerization and growth
- Experiences wet–dry cycles (microbial spores, seeds of higher plants, tardigrades)
- Can thrive in elevated temperature ranges
- Employs catalysts and metabolic pathways
- Uses light as a primary source of energy
- Requires genetic information
- Is based on cellular compartments
- Grows by polymerization
- Uses feedback regulation to control life processes
- Divides into daughter cells
- Exists as populations of cells (microbial mats, tissues)
- Evolves over time toward ever increasing complexity

Here is the question: *How could such complex systems begin on a sterile early Earth?*

Many more details need to be worked out in a continuing research effort before we can reach a consensus about the physical and chemical environment of the prebiotic Earth from which life emerged. To this end, I will explore what we know with a reasonable degree of confidence, then point the way toward the next level in which seemingly unrelated details could coalesce into a comprehensive map that will let us understand how life can begin on a habitable planet.

References

Deamer D (2017) Conjecture and hypothesis: The importance of reality checks. *Beilstein J Org Chem* 13, 620–624.

Platt JR (1964) Strong Inference. *Science* 146, 347–353.

The Early Earth

An Ocean with Volcanoes

> If we had a time machine and could travel to the ancient Pilbara we would see great volcanoes emerging from the sea, and small areas of other land, but no continent. It would have looked like the Hawaiian Islands in the middle of the Pacific Ocean, but with little sign of life: no trees, no grass, no animals. No surfers on the beach.
>
> Malcolm Walter

Overview and Questions to be Addressed

Malcolm Walter was talking about the Pilbara region of Western Australia where some of the oldest known biosignatures of ancient life, in the form of extensive fossilized stromatolites, are preserved. The first potential stromatolite was discovered by graduate student John Dunlop, who was studying barite deposits at the North Pole Dome. Roger Buick went on to investigate the biogenicity of the stromatolites for his PhD (Buick, 1985) and Dunlop, Buick, and Walter published their results (Walter et al., 1980). In a prescient paper, Walter and Des Marais (1993) proposed that the ancient stromatolite fossils could guide the search for life on Mars. I have walked with Malcolm Walter through the Dresser formation where the fossils were found. It is humbling to realize that if time passed at a thousand years per second, it would take 41 days to go back in time to the first signs of life on our planet.

In any description of events that occurred some 4 billion years ago, certain assumptions must be made. I will try to make assumptions explicit throughout this book, beginning here with the geochemical and geophysical conditions prevailing on the early Earth and Mars. I am including Mars not as an afterthought but because both planets had liquid water 4 billion years ago. Most of our understanding of planetary evolution comes from observations of our own planet, but it is now clear that the Earth and Mars were undergoing similar geophysical processes during the first billion years of the solar system's existence, with an equal probability that life could begin on either planet. In a sense, the surface of Mars is a geological fossil that has preserved evidence of what was happening there at the same time that life began on the Earth. For instance, Martian volcanoes offer direct, observable evidence that volcanism was occurring nearly 4 billion years ago; making it plausible that similar volcanism was common on Earth even though the evidence has been completely erased by geological and tectonic processes.

I have often noticed that papers related to the origin of life tend to assume that water is simply a liquid solvent with little attention paid to its properties. The main point to be made in this chapter is that if water is essential for the origin of life, it is important to understand the source of the water, the time at which it first appeared, the geophysical environment of the first hydrosphere, and the chemical composition of the water. There are two primary aqueous bulk phases today and a reasonable assumption is that this was also true 4 billion years ago. A major fraction would be salty seawater in the original global ocean but a smaller fraction was present as water distilled from the ocean and precipitated on volcanic islands to form freshwater sites characterized by geysers, hot springs, and associated hydrothermal fields.

Questions to be addressed in Chapter 1:

- Where did the Earth's water come from?
- When did the ocean appear?
- How salty was the ocean?
- Was there also freshwater on the prebiotic Earth?

Sources of the Earth's Water

In one way or another, I have been associated with NASA-sponsored research since the 1980s when I served on the peer review panel for the exobiology program. As I watched exobiology grow and evolve into astrobiology, with the expectation that we might find life on other planets, there was a central rule: Look for liquid water.

Even though water is integral to life on Earth, there is a surprising degree of uncertainty about its source. There is no doubt that water is a common constituent of the interstellar medium because its characteristic spectrum is readily distinguished in the microwave region when radio astronomers observe the molecular clouds that give rise to stars and solar systems (Irving, 1991). Much of the water is present as thin layers of mixed ice on interstellar dust particles, along with ammonia, carbon monoxide, and methanol (Allamandola et al., 1999). Therefore, as gravity caused dust particles in the original solar nebula to aggregate, the water ice was mixed with a variety of carbon compounds and integrated into planetesimals and comets, then delivered to planetary surfaces during the accretion process by which planets like the Earth and Mars formed.

The story begins with the final phase of accretion that took place after ~90% of the Earth's mass was already present. During this phase, a Mars-sized planet collided with the proto-Earth to add the last 10% of the Earth's mass (Canup and Asphaug, 2001). The violent collision produced a splash of molten rock and vapor that formed a ring around the Earth resembling that of Saturn, and this material gathered by gravitational forces into the moon. This idea was recently challenged, because the compositions of the lunar and terrestrial minerals are very similar. Instead of a single collision between planets, perhaps a series of smaller objects impacted the Earth, mixing into the surface and also splashing material into a disk from which the moon formed (Rufu et al., 2017).

Whichever scenario is correct does not affect the narrative being presented in this book, because in either case so much energy was released that the surfaces of the Earth and moon were magma oceans at the temperature of molten volcanic lava. The energy of impacts also eroded most of the volatile compounds in the original atmosphere, leaving behind a steam atmosphere with significant components of nitrogen gas and carbon dioxide.

An important point is that no organic compounds could have survived this temperature range, so all of the organics required for life to begin must have become available only after the Earth cooled sufficiently for an ocean to condense from an atmosphere of water vapor. There are two plausible sources that will be discussed in detail in Chapter 4. The required organic material could have been synthesized at the Earth's surface by geochemical and photochemical reactions or delivered during late accretion as meteoritic infall and comets. Most likely both sources contributed to the suite of organic compounds, but in either case, the compounds would undergo processing by secondary chemical reactions at the Earth's surface and atmosphere before being caught up in the chemical evolution leading to life.

There are also two possible sources of the Earth's water: comets impacting the cooling Earth after the moon-forming phase and volcanic outgassing from the Earth's interior. Comets are at one end of the mineral-water spectrum, typically composed of water ice mixed with smaller amounts of mineral dust. Almost certainly comets did contribute some of the Earth's water, but they were probably not the primary source. Some insight comes from comparisons of the isotopic composition of seawater and cometary ice. There are two general classes of comets, depending on whether they originated in the Kuiper Belt or Oort Cloud (Mumma and Charnley, 2011). The deuterium/hydrogen (D/H) ratio of a few comets has been determined, and this can be compared to terrestrial seawater. If the D/H ratios match those of comets, a reasonable conclusion would be that the ocean is largely cometary in origin. However, the initial indications are that with one exception, the D/H ratios of comets do not perfectly match that of seawater but instead match the D/H ratios of the water in carbonaceous meteorites. The tentative conclusion is that the major source of the Earth's water was not delivered by comets but by impacting planetesimals during primary accretion. This is supported by recent evidence.

Previous measurements of the deuterium/hydrogen (D/H) ratio in other comets have shown a wide range of values. Of the 11 comets for which measurements have been made, it is only the Jupiter-family Comet 103P/Hartley 2 that was found to match the composition of Earth's water in observations made by ESA's Herschel mission in 2011.

By contrast, meteorites originally hailing from asteroids in the Asteroid Belt also match the composition of Earth's water. Thus, despite the fact that asteroids have a much lower overall water content, impacts by a large number of them could still have resulted in Earth's oceans.

It is against this backdrop that Rosetta's investigations are important. Interestingly, the D/H ratio measured by the Rosetta Orbiter Spectrometer for Ion and Neutral Analysis, or ROSINA, is more than three times greater than for Earth's oceans and for its Jupiter-family companion, Comet Hartley 2.

Indeed, it is even higher than measured for any Oort cloud comet as well. (Rosetta mission)

The planetesimals impacting the Earth during the accretion phase were composed of a variety of stony minerals and metallic elements mixed with water ice and organic compounds. A tiny fraction of planetesimals escaped planet formation and are still orbiting the sun as asteroids between Mars and Jupiter, ranging in size from tens of meters to a few like Ceres and Vesta that have diameters of hundreds of kilometers. Smaller asteroids are loose aggregates composed of dust and sand-sized minerals mixed with small amounts of water ice. The interiors of larger asteroids were heated by radioactive decay, getting hot enough to melt iron and nickel and undergoing differentiation as the molten metals sank into a core, leaving behind a rocky crust. Asteroids are constantly colliding with one another, and as a result of the impacts, smaller pieces are ejected into the surrounding space. Most of these simply continue to orbit within the belt, but a rare few fall sunward and intersect the Earth's orbit. If they survive the fiery trip through the atmosphere and fall to the Earth's surface, they are called meteorites.

Because asteroids assembled very early in solar system history, approximately 4.6 billion years ago, meteorites derived from collisions between asteroids provide the richest source of information we have about the composition of the asteroid parent bodies. Most meteorites are stony, a few are metallic, composed of a nickel-iron alloy, and a small fraction are carbonaceous, containing a complex suite of organic compounds. This is the most convincing evidence that biologically relevant compounds can be synthesized by nonbiological processes. Later in the book we will use carbonaceous meteorites as a guide to the kinds of organic compounds that were plausibly available in the prebiotic environment of our planet.

It is certain that all of the biogenic elements—carbon, oxygen, nitrogen, phosphorus, and sulfur—now present on the Earth were delivered by planetesimals during primary accretion. Along with water, delivery included the silicate minerals of the mantle and crust, and the iron-nickel mixture that sank to the core as molten metal. Compounds related to the origin of life were certainly present, along with silicate minerals and metallic iron and nickel. However, none of the more complex organic molecules could survive the temperatures of primary accretion and the moon-forming event, so the composition of the Earth's surface would simply be the most stable compounds that survived the process, including molecular nitrogen (N_2) and carbon dioxide in the atmosphere.

The First Liquid Water

Even though the ocean seems vast, in fact it is a very thin layer of water coating a mass of rocky mineral crust. If the Earth were the size of a basketball 25 cm in diameter, the ocean would resemble a 75-micrometer layer of dew, about the thickness of a sheet of paper. It does not require much water to produce an ocean.

We can use what we know about the history of our planet to make a reasonable estimate of when the first ocean formed. Microscopic fossils of early life have been discovered in Australian and South African rocks that are ~3.5 billion years old, and the origin of the Earth itself can be dated to ~4.6 billion years ago based on the known half-life of uranium isotopes in meteorites. Another clue comes from the elemental composition of zircons, which is related to the temperature at which they formed. The surprising discovery is that the temperature was much lower than that of magma (Wilde et al., 2001) indicating that the Earth's crust was cool enough for liquid water to exist in the form of a global ocean approximately 4.4 billion years ago. There is no way to know how long it took for the first primitive forms of life to appear in the billion-year interval between the ocean condensing and the 3.5-billion-year-old microfossils and stromatolites, so we will choose 4 billion years as a point in time when the events described in this book were occurring, with an uncertainty of perhaps 250 million years on either side of that number.

If we agree that water is essential for life, the history of life on Earth began when the first liquid water fell to the hot surface of a cooling magma ocean. It seems likely that this would have been in complete darkness because the Earth was shrouded in the global cloud of a steam atmosphere. The first drops reaching the magma exploded back into vapor, but the basaltic surface became slightly cooler as a result. The heat absorbed by the water was transported into the atmosphere and eventually radiated into space. The first few drops were followed by vast numbers of others, and at some point, a thin film of liquid water, the start of an ocean, remained in cracks and depressions in the lava. The temperature was much higher than the 100° C of boiling water today because the immense atmospheric pressure of an entire ocean of water was still present as vapor. Zahnle et al. (2010) estimated that the atmospheric pressure at the end of accretion was in the range of 270 bar, of which 140 bar was water delivered by impacting planetesimals and 100 bar was the result of outgassing. (One bar is defined as the atmospheric pressure of the Earth today.)

As the Earth cooled further, radiating heat into outer space, the thin film of water slowly thickened, finally approaching the depth of today's ocean. (The average depth now is 4237 meters.) During this process the atmospheric pressure fell, and the cloud cover thinned, allowing a more rapid cooling of the surface. At some point the first sunlight broke through the clouds and illuminated a violent scene. Volcanoes had emerged from the ocean and were spewing immense amounts of ashes into the atmosphere, along with water vapor, carbon dioxide, hydrogen chloride, and sulfur compounds (Plate 1.1). The first volcanoes on Earth were lost to erosion long ago, but ancient volcanoes have been preserved on Mars. For instance, Apollinaris Mons has been dated to 3.9 Gya, and the much younger (~100 million years ago) Olympus Mons rises 25 kilometers above the surrounding plane, twice the height of the Hawaiian volcanoes at just 10 kilometers above their surrounding sea floor. Because the moon was much closer to the Earth, it is probable that vast tides hundreds of meters high washed over the shores of the first volcanic islands.

Most important for the narrative in this book, water vapor was continuously being distilled from the ocean by evaporation and precipitated as endless rainfall on the slopes of the volcanoes. At first the freshwater simply disappeared into the volcanic ash and crevices in the lava, flowing downhill back to the sea as it does on Hawaii today. However, over millions of years the action of heat and acidic water chemically transformed the ash and lava surfaces from basalt to aluminum silicate mineral clay. The microscopic clay particles began to seal the crevices and retain water, producing fluctuating pools fed by geysers, hot springs, and cyclic precipitation. In this book, such sites will be referred to as hydrothermal fields similar to those sites that are abundant in the volcanic regions of today's Earth.

The Early Atmosphere

A reasonable assumption is that the early atmosphere was the result of outgassing from the interior of the Earth following the moon-forming event. Trail et al. (2011) noted that the composition of the Hadean atmosphere depended on the fugacity of the gas mixture as it passed through hot magma on the way to the surface. Fugacity is a term used to define the composition and chemical behavior of real systems (as opposed to ideal systems) and is expressed in units of pressure. It is also related to the chemical potential of a reacting system, not as an actual property of matter but instead calculated from measurements. Trail et al. measured the cerium composition of zircon crystals known to have an age in the range of ~4 Gya, which means that they were produced in the Hadean. Cerium is a lanthanide metal that exists either in a +3 or +4 oxidation state. By measuring the oxidation state of the cerium in the zircon, the authors concluded that the atmosphere of the Hadean could not have been in a reduced state as hydrogen, methane, and ammonia, nor would free oxygen have been a major component. Instead, the most likely atmosphere 4 billion years ago would be composed of nitrogen (N_2), carbon dioxide (CO_2), sulfur dioxide (SO_2), and water vapor (H_2O). For the purposes of this book, I will use this composition to guide simulations that incorporate a gas assumed to reflect the composition of the prebiotic atmosphere.

A Geology Lesson

Four words—ultramafic, mafic, andesitic, and felsic—often appear in geological research literature to describe the lavas and ashes that are produced when magma erupts from a volcano. The words basically refer to the temperature at which the minerals were formed and the relative composition of silica- to iron- and magnesium-rich minerals. Ultramafic lava forms at high temperatures up to 1500° C, has less than 45% silicate, and a relatively high content of iron and magnesium (8–32%). Ultramafic minerals are present in Archean lavas but not in volcanic eruptions today.

Mafic lavas are composed of basalt cooling to a solid at a temperature of ~1300° C and have a silicate content of ~50% with iron and magnesium content less than 10%. Common examples of mafic lavas are those produced at divergent plate boundaries (also known as spreading centers) such as those found in the mid-Atlantic and Pacific Oceans and referred to as mid-ocean ridges (MOR). These are associated with deep-sea hydrothermal vents (black smokers). Mafic lavas are also associated with hot spot volcanism like that in Hawaii and Iceland. In the case of Iceland, however, hot spot volcanism is combined with a spreading center since Iceland also sits along an MOR. Another hot spot is Yellowstone National Park, but this is under a continent, so the hydrothermal activity is in the form of hot springs rather than deep sea vents. Andesitic lavas are produced from magma at 1000° C and have ~60% silicate content and relatively low iron and magnesium (~3%). Finally, felsic lavas are associated with the lowest temperature (900° C), highest silicate (70%), and lowest iron and magnesium (2%). Felsic igneous rocks occur in the volcanoes emerging from subduction zones along continental margins. Subduction occurs when heavier oceanic crust spreading from the MOR is forced under the lighter continental crust. The addition of ocean water during subduction and the melting of the overlying continental crust produces felsic rock enriched in silicate minerals that melt into volcanic lava.

The reason it is important to understand the chemical composition and physical properties of volcanic minerals is that these characteristics affect the way that water and dissolved solutes behave when they come in contact with different types of volcanic rocks. There is an important secondary reaction of lava with water that involves aluminum, which at 8.1% by weight is the most abundant metallic element in the Earth's crust, followed by iron (5%), calcium (3.1%), sodium (2.8%), potassium (2.6%), and magnesium (2.1%). All of these metals are soluble in their ionized form and dissolve in water at varying rates. After millions of years of exposure to acidic water and elevated temperatures in hydrothermal fields, lava is slowly altered to minerals called clays. These are aluminum phyllosilicates characterized by the small size of the crystals (<2 micrometers) and a layered structure of sheets. The structure and properties of clays have long been a focus of research into the origin of life, as described in Chapter 5.

How Hot and Salty Was the First Ocean?

We don't know the temperature and salt content of the early ocean with certainty, but it is an important aspect related to the origin of cellular life, so we need to make some educated guesses. Paul Knauth (2005) used oxygen isotope data to estimate temperature and also estimated the salt content of the early ocean from known values of voluminous salt deposits produced by evaporating seas widespread around the Earth's crust. He writes:

> The temperature and salinity histories of the oceans are major environmental variables relevant to the course of microbial evolution in the Precambrian, the "age of microbes". Oxygen isotope data for early diagenetic cherts indicate surface temperatures on the order of 55–85 °C throughout the Archean, so early thermophilic microbes (as deduced from the rRNA tree) could have been global and not just huddled around hydrothermal vents as often assumed. Initial salinity of the oceans was 1.5–2× the modern value and remained high throughout the Archean in the absence of long-lived continental cratons required to sequester giant halite beds and brine derived from evaporating seawater.

Knauth's estimates are reasonable approximations of conditions in Earth's earliest ocean at the time that life began. Other important solutes are divalent cations such as calcium, magnesium, and ferrous iron, but less attention has been paid to these. Given that the initial pH of the ocean was in the acidic range (due to atmospheric carbon dioxide and hydrochloric acid released by volcanoes) and that there was violent agitation of the crust as the ocean formed, it seems reasonable that divalent cations were also dissolved from mineral surfaces and were present in ocean water just as they are today.

We can make rough "back of the envelope" estimates of the amounts by assuming that the calcium, magnesium, and iron in sedimentary minerals were once dissolved in the ocean, and then calculate their concentration. During the first 2 billion years of the Earth's history, there were only trace amounts of molecular oxygen present. The strongest evidence is in the mineral record, which reveals a shift to oxidized minerals that began around 2.5 billion years ago. As oxygen produced from photosynthetic microorganisms called cyanobacteria began to accumulate in the ocean, it reacted with soluble ferrous iron (Fe^{2+}) resulting in ferric oxide precipitation as hematite and magnetite (Fe_2O_3). Because these are valuable iron ores, their global abundance has been carefully estimated and totals approximately 90 billion metric tons, equivalent to 30 billion tons of elemental iron. When this is converted to moles of iron originally dissolved in the volume of the ocean, it turns out that the ferrous iron concentration 4 billion years ago would only be in the range of 2 micromolar. Today's value is 0.03 micromolar, so most of the original soluble ferrous iron in the ocean has been oxidized to its ferric form and precipitated in vast deposits we now refer to as Banded Iron Formations (BIFs), which can host economically viable iron ore.

We can do a similar calculation for calcium concentration. The amount of carbon in the Earth's crust as carbonate has been estimated to be 48×10^{21} g, most of which is $CaCO_3$. This is equivalent to 0.48×10^{21} moles of calcium, and dividing by the volume of the ocean in liters gives a calcium concentration of 0.34 M. This is absurdly high because much of the mass of calcium in sedimentary limestone today was delivered to the ocean by erosion of land masses after continents emerged. On the other hand, if only 3% of the total calcium was dissolved in prebiotic seawater, that would match the 10 mM Ca^{2+} concentration in today's ocean, so we will assume that the divalent Ca^{2+} concentration was similar to contemporary values. This is an important assumption, because calcium and magnesium are divalent cations that markedly affect the self-assembly processes to be described in later chapters.

Was Freshwater Available at the Time of Life's Origin?

Even though most of the water on the early Earth was in the ocean with dissolved salts, there were likely to be sources of freshwater as well. The freshwater produced by evaporation from the salty ocean would condense and precipitate as distilled rain water on volcanic land masses. Today, only 1% of the Earth's water is fresh, mostly in the form of massive ice fields in Antarctica and Greenland. Although smaller amounts are present in the Great Lakes and Lake Baikal in eastern Russia, fresh rain water is continuously inundating volcanic islands like Hawaii. In fact, knowing that the annual rainfall in the Hawaiian islands is about one meter, and the area of the islands is 16,600 km², the total annual volume of rain is 16.6 km³. The ocean has a volume of 1.4×10^9 km³, so the equivalent of an entire ocean falls on Hawaii every 100 million years! Assuming that there were multiple volcanic islands the size of Hawaii on the early Earth with similar precipitation rates, a reasonable conclusion is that there would be sufficient freshwater available for the kinds of natural experiments leading to life that are described in later chapters. In other words, there is no reason to believe that salty seawater was the only possible site for life to begin.

Fossil Evidence for the First Microbial Life

This book is about the steps leading to the origin of life, so we will not spend much time on the evidence related to the first appearance of microbial life, other than to use it as a time point when the preliminary steps have been completed. A timeline illustrating the entire period that life has existed on the Earth is presented in Plate 1.2

As described at the beginning of this chapter, the most conspicuous evidence of early life are fossil stromatolites, first discovered in the late 1970s, in the Pilbara region of Western Australia (Walter et al., 1980; Buick et al., 1981). The term stromatolite comes from Greek words meaning layered rock. Stromatolites are produced when microbial mats become mineralized as they grow on rock surfaces. (Familiar structures that resemble stromatolites are the plaques that form on teeth and are removed by visits to a dentist. The plaques are composed of layers of a calcium phosphate mineral called apatite that is deposited by films of bacteria.) The stromatolite fossils in the Pilbara were originally assumed to have formed in a marine estuary similar to Shark Bay on the coast of Western Australia, but more recent work has re-interpreted the environment as an ancient volcano (Van Kranendonk et al., 2008) with reports of geyserite and sinter deposits by Djokic et al. (2017) indicating that this volcanic system once hosted land-based, freshwater hot springs like those in Yellowstone National Park. Evidence for life living in and around these hot springs includes preserved bubbles inferred to have been trapped in a sticky microbial mucous, and palisade fabric that represents entombed microbial filaments. We will return to this discovery in later chapters because it is relevant to the question of whether life began in salty seawater at deep sea hydrothermal vents or freshwater associated with hydrothermal fields.

Further evidence that microorganisms were present in a terrestrial freshwater environment came from measurements of sulfur isotope distribution in pyrite (iron sulfide). The pyrite crystals exhibited clear evidence that they had been in what is called a paleosol, or ancient soil, in a river plain approximately 3.2 billion years ago. Furthermore, the surfaces of the crystals had been modified in a way characteristic of microbial growth. "This indicates that vadose-zone soil-forming processes in the Archean involved not only physical and chemical modification of moist, unconsolidated sediment in a terrestrial environment but also already included its microbiological modification" Nabhan et al. (2016).

Summary: Setting the Stage

- The steps toward the origin of life began approximately 4 billion years ago after atmospheric water vapor had cooled sufficiently to condense into a global ocean.
- The atmosphere was a mixture of nitrogen gas, carbon dioxide, and water vapor.

- The presence of liquid water provided a medium required for the physical and chemical processes leading to the origin of life.
- Although most of the Earth's water was in the ocean, volcanism produced island land masses on which precipitation fed hydrothermal fields consisting of freshwater cycling within hot springs, geysers, and evaporating pools.
- Contemporary analogues of such sites include the islands of Hawaii and Iceland.
- Volcanism also produced submarine hydrothermal vents, but in the absence of continents and plate tectonics these more closely resembled the hydrothermal activity of sea mounts over magma plumes.
- The global temperature was elevated, so the first steps toward life occurred at temperature ranges associated with hydrothermal fields today.
- The water circulating in hydrothermal fields was acidic due to hydrochloric acid (HCl) and sulfur compounds present in volcanic vapor.

References

Allamandola L, Bernstein, MP, Sandford SA (1999) Evolution of Interstellar Ices. *Space Sci Rev* 90, 219–232.

Buick R (1985) Life and conditions in the early Archaean: Evidence from 3500 MY old shallow-water sediments in the Warrawoona Group, North Pole, Western Australia. University of Western Australia

Buick R, Dunlop JSR, Groves DI (1981) Stromatolite recognition in ancient rocks: An appraisal of irregularly laminated structures in an early archaean chert-barite unit from North Pole, Western Australia. *Alcheringa* 5, 161–181.

Canup RM, Asphaug E (2001) Origin of the Moon in a giant impact near the end of the Earth's formation. *Nature* 412, 708–712.

Djokic T, Van Kranendonk MJ, Campbell KA, Walter MR, Ward CR (2017) Earliest signs of life on land preserved in ca. 3.5 Gya hot spring deposits. *Nat Commun* 8, 15263. doi: 10.1038/ncomms15263.

Irving W (1991) Chemical abundances in cold, dark interstellar clouds. *Icarus* 91, 2–6.

Knauth LP (2005) Temperature and salinity history of the Precambrian ocean: implications for the course of microbial evolution. *Palaeogeography, Palaeoclimatology, Palaeoecology* 219, 53–69.

Mumma MJ, Charnley SB (2011) Chemical composition of comets: Emerging taxonomies and natal heritage. *Ann Rev Astron Astrophys* 49, 471–524.

Nabhan S, Wiedenbeck M, Milke R, Heubeck C (2016) Biogenic overgrowth on detrital pyrite in ca. 3.2 Ga Archean paleosols. *Geology* 44, doi: 10.1130/G38090.1.

Rosetta mission http://www.esa.int/Our Activities/SpaceScience/Rosetta/Rosetta fuels debate on origin of Earths oceans.

Rufu R, Aharonson O (2017) A multiple-impact origin for the Moon. *Nat Geosci* 10, 89–94.

Schopf JW (1993) Microfossils of the early Archean Apex Chert: New evidence of the antiquity of life. *Science* 260, 640–646.

Trail D, Watson EB, Ailby ND (2011) The oxidation state of Hadean magmas and implications for early Earth's atmosphere. *Nature* 480, 79–82.

Van Kranendonk MJ, Philippot P, Lepot K, Bodorkos S, Pirajno, F (2008) Geological setting of earth's oldest fossils in the ca. 3.5 Ga Dresser formation, Pilbara craton, Western Australia. *Precamb Res* 167, 93–124.

Wacey D, Kilburn MR, Saunders M, Cliff J, Brasier MD (2011) Microfossils of sulphur-metabolizing cells in 3.4-billion-year-old rocks of Western Australia. *Nat Geosci* 4, 698–702.

Walter MR, Buick R, Dunlop JSR (1980). Stromatolites 3,400–3,500 myr old from the North Pole area, Western Australia. *Nature* 284, 443–445.

Walter M, Des Marais DJ (1993) Preservation of biological information in thermal spring deposits: developing a strategy for the search for fossil life on Mars. *Icarus* 101, 129–143.

Wilde SA, Valley JW, Peck WH, Graham CM (2001) Evidence from detrital zircons for the existence of continental crust and oceans on the Earth 4.4 Gyr ago. *Nature* 409, 175–178

Zahnle K, Schaefer L, Fegley B (2010) Earth's Earliest Atmospheres. *Cold Spring Harbor Perps Biol.* doi: 10.1101/cshperspect.a004895.

2

Geochemical and Geophysical Constraints on Life's Origin

Life is a partial, continuous, progressive, multiform and conditionally
interactive self-realization of the potentialities of atomic electron states.
J.D. Bernal (1967)

Overview and Questions to Be Addressed

Bernal's quote is a bit wordy, but he was basically saying that life can be understood as a continuous chemical reaction, and I agree. Throughout this book I will
be describing ideas about how life can begin on habitable planets, which are defined
as planets with orbits not too close and not too far from a star so that the temperature permits liquid water to exist. The conditions in which life can begin must have
sufficient complexity to permit primitive life to assemble from organic chemicals
dispersed in a sterile environment which then begin to react and evolve into more
complex structures. This chapter will describe the main parameters of geochemical
and geophysical complexity, and then consider them in terms of scales from the
nanoscopic to the macroscopic.
    Questions to be addressed:

- What scales must be considered to understand how life can begin?
- What are the properties of the scales?
- How do the scales relate to the origin of life?

Prebiotic Earth and Mars: Global Scales

The physical dimensions related to the origin of life can be described in terms of four
scales—global, local, microscopic, and nanoscopic—and these dimensions must be related to the chemical and physical properties of each scale. The global scale is easiest to
understand because the parameters are averages of very broad variables. For instance,
we can state that the global temperature today is 15° C and even follow changes in the
temperature to accuracies of a tenth of a degree on a year to year basis. However, within
the global scale are extreme variations between winter temperatures of -60° C at the

Table 2.1 **Global Parameters of Earth and Mars**

Parameter	Earth today	Earth 4 Gya	Mars today	Mars 4 Gya
Mass	5.98×10^{24} kg	Same	6.4×10^{23} kg	Same
Radius	6370 km	Same	3394 km	Same
Ocean mass/ volume	1.4×10^{21} kg or liters	Similar	None	$\sim 5 \times 10^{18}$ kg (est.)
Surface area	510 million km^2	Same	145 million km^2	Same
Continental surface area	149 million km^2	None	None	None
Atmosphere mass	5.15×10^{18} kg	Similar	2.5×10^{16} kg	
Atmospheric gases	78% N$_2$ 21% O$_2$ 1% Ar 0.04% CO$_2$	Mostly N$_2$ Some CO$_2$ Trace O$_2$	95% CO$_2$ 3% N$_2$ 1.6% Ar 0.13% O$_2$	Uncertain
Global temperature	15° C	55–85° C	–57° C	>0° C
Ocean pH	8.1	5–6 (acidic)	N.A.	
Biomass	5.6×10^{14} kg C	None	None	Uncertain
Total organic carbon	1.2×10^{15} kg			
Total carbon in crust	4.7×10^{21} kg	Similar		

poles and summer temperatures of 50° C in Death Valley, California. Of course, even higher temperatures are associated with hydrothermal fields, up to boiling at 100° C, but sometimes nearer to 90° C because the fields are usually at higher elevations associated with volcanoes. Table 2.1 summarizes the main parameters of the global scale on Earth and Mars today and compares their values with those near the time that life began on the Earth 4 billion years ago.

Local Parameters

Although global scales are relevant to our understanding of how life could begin, even more important are localized sites with dimensions measured in meters to hundreds of meters. This means that we will pay attention to the way that water interacts with mineral surfaces and the atmosphere and in particular, to the physical and chemical interactions of organic solutes under these conditions.

Familiar examples of local scales on today's Earth include hydrothermal vents—black and white smokers—and hydrothermal fields like those in Yellowstone National Park, Iceland, New Zealand, and the volcanoes of Kamchatka in eastern Russia. We can learn from such sites because they are analogous to the hydrothermal conditions 4 billion years ago when life began on the Earth and perhaps on Mars. One of the distinguishing characteristics of vents and fields is the complexity of interfaces found there. Vents have a single interface between minerals and seawater, while fields have three interfaces that undergo continuous fluctuations: mineral/water, mineral/atmosphere, and atmosphere/water. As will be discussed in later chapters, the physical and chemical processes that occur at fluctuating interfaces are an essential part of the hypothetical scenario being developed in this book.

The second distinguishing difference between vents and fields concerns ionic solutes in the water. Seawater has a relatively high concentration of sodium chloride and divalent cations like calcium and magnesium, and these strongly affect the self-assembly of amphiphilic compounds into membranous compartments. In contrast, the water of hydrothermal fields is essentially distilled water produced by evaporation of seawater and precipitated as rain on volcanic land masses. The acidity and alkalinity of the water are also important. Some hydrothermal fields are acidic (pH 2–4) due to the SO_2 and HCl vapor that dissolves in the water, while others are in the neutral pH range. The aqueous fluids of hydrothermal vents have a much broader pH range from pH 3 for black smokers to pH 10–11 for alkaline vents. In a later chapter, we will discuss how pH strongly affects the kinds of chemical reactions that can occur between organic solutes dissolved in water.

Finally, we need to pay attention to the composition of minerals that come into contact with hydrothermal water. The primary mineral surface associated with volcanic hydrothermal fields is basalt, but over thousands to millions of years of exposure to acidic hydrothermal water, the basalt slowly weathers to become microscopic particles of various clays. The adsorptive properties of clay interfaces have been the focus of many earlier studies, but here the focus will be on the physical properties of clays as sealants of otherwise porous volcanic minerals.

The minerals of hydrothermal vents have significant roles in the speculations that life could emerge in the chimneys of black or white smokers. Particular emphasis has been given to the fact that black smokers have abundant iron sulfides and that iron-sulfur complexes are present in at least 16 enzymes, such as hydrogen dehydrogenase, carbon monoxide dehydrogenase, and the ferredoxin in nitrate reductase and glutamate synthetase. It is possible to imagine that such enzymes were active in primitive forms of microbial life (Russell and Martin, 2002).

Hydrothermal fields are a plausible alternative to hydrothermal vents as sites for the origin of life. The measured pH of hydrothermal water accumulating in basaltic mineral basins is low due to acidic components of volcanic vapor that dissolve in the water. This acidic pH is significant because it is related to the availability of phosphate which is limited by the fact that when calcium and phosphate are mixed in neutral to alkaline pH ranges, a highly insoluble mineral called apatite is formed. However, the acidic water of hydrothermal fields would readily dissolve apatite and make soluble phosphate available for phosphorylation of organic compounds, an essential first step toward primitive metabolism.

Microscopic Parameters: Properties of Minerals

The third scale to consider has dimensions measured in micrometers and can be investigated using light microscopy. Glass containers are commonly used in laboratory simulations of prebiotic conditions, and microscopically, glass surfaces appear entirely smooth. This is acceptable if the experiments are being done in solution, but with the possible exception of obsidian and mica there is no equivalent of smooth glass in natural conditions. Instead the most common mineral surfaces on the Earth 4 Gya would be volcanic ash, cooled lava, and pumice. However, these are dry surfaces, and the sites we are considering as analogues for the origin of life are today's hydrothermal fields. Plate 2.1 shows small, liter-sized pools on Mount Lassen in northern California and Mount Mutnovsky in Kamchatka that we are simulating in the laboratory. Both of these sites undergo cyclic fluctuations of water level due to precipitation and hot spring activity, and both have layers of clay that seal the underlying lava and ash so that water can accumulate as transient pools. The pool on the right in the plate is evaporating, and "bathtub rings" of dried films can be seen around the edges of rocks. The point is that clay and silicate are the two most common mineral species in hydrothermal ponds today and are presumably representative of the kinds of mineral surfaces available in the prebiotic environment.

Clay minerals have been a focus of interest in origins of life research ever since Bernal (1951) suggested that they might play an important role. The crystals are microscopic, with dimensions in the micrometer size range (Fig. 2.1). As a result, they have extraordinarily large surface areas: the estimated areas of one gram of two different clays range from 100 to 700 square meters! The surface area has both positive and negative implications for the origin of life. The positive effect is that the clay mineral surface is electrically charged. As a result, clay minerals have a remarkable capacity to adsorb organic solutes which then become concentrated on the surfaces. Graham Cairns-Smith (2005) speculated that the nanoscopic organization of polar and ionic groups on clay surfaces could have served as a mineral version of genes so that specific organic compounds could line up on the surface and perhaps polymerize into a linear strand resembling a nucleic acid. Jim Ferris (2002) and his colleagues found that activated nucleotides were in fact adsorbed to Montmorillonite clay surfaces, with the result that oligomers up to 15 or more nucleotides in length could be synthesized from activated adenosine monophosphate. This result will be described in detail in Chapter 5 on self-assembly.

The negative charge on clay mineral surfaces contributes to their immense adsorptive capacity. The fact that solutes are strongly adsorbed also means that they are being removed from solution as active reactants, thereby reducing the chances that they can participate in chemical evolution on the path toward life (Deamer et al., 2006). Furthermore, there is no guarantee that they will undergo useful reactions if they are bound to clays because they are no longer freely diffusing. Even if they do react, the products may be so tightly bound to the clay that they cannot escape back into solution.

Despite the interest in mineral surfaces, their chemical characteristics turn out to be not very significant to the narrative being developed in this book. Here the emphasis is related to the biophysical properties of membranous vesicles composed of amphiphilic

Figure 2.1. This electron microscopic image of clay minerals reveals its layered structure. Bar shows 10 micrometers. Photograph by Evelyne Delbos, the James Hutton Institute.

compounds. As will be discussed in later chapters, when amphiphilic compounds are included in the mixture of potential monomers, followed by evaporative dehydration, the amphiphiles coat the mineral surfaces with multilayered structures composed of the amphiphiles, with the reactants captured between the layers. This means that most of the reactants never come into close contact with the minerals, but instead are dissolved in protective layers of amphiphilic molecules. The primary role of the minerals has to do with their contribution of ionic solutes to the aqueous phase and their buffering effect on the pH of the medium rather than any specific property of the minerals themselves. *This is an absolute difference between the scenario proposed here and other scenarios that involve solution chemistry in water or adsorption to mineral surfaces.*

Nanoscopic Parameters

Self-assembly is central to all life today and very likely played an essential role in the processes leading to the origin of life. For example, amphiphilic compounds like phospholipids assemble spontaneously into lipid bilayers that provide the boundaries of all cellular life. Other examples include peptide chains folding into functional configurations and nucleotide subunits of nucleic acids undergoing base pairing and base stacking. These processes occur at the nanoscopic level and are so significant to the narrative being developed in this book that Chapter 5 is devoted to a detailed discussion of their properties.

Setting the Stage

Depending on the disciplinary affiliation of researchers who study the origins of life, there is a tendency to think about the problem at four different scales. Planetary scientists carry out research on global scales, geologists and ecologists on local scales, cell biologists on microscales, and biochemists on nanoscales. However, in order to understand how life can begin, the scales must be integrated in terms of components, structure, physical and chemical properties, energy flow, and interactions across dimensions that range over 12 orders of magnitude, from nanometers to thousands of kilometers. Plate 2.2 illustrates this range, and this book attempts to provide an integrated perspective that incorporates all four scales.

References

Bernal JD (1951) *The Physical Basis of Life*. London: Routledge & Kegan Paul.

Cairns-Smith AG (2005) Sketches for a mineral genetic material. *Elements*. doi: 10.2113/gselements.1.3.157.

Deamer D, Singaram S, Rajamani S, Kompanichenko V, Guggenheim S (2006) Self-assembly processes in the prebiotic environment. *Phil Trans R Soc Lond B* 361, 1809–1818.

Ferris JP (2002) Montmorillonite catalysis of 30–50mer oligonucleotides: Laboratory demonstration of potential steps in the Origin of the RNA World. *Orig Life Evol Biosph* 32, 311–332.

Russell MJ, Martin W (2002) The rocky roots of the acetyl-CoA pathway. *Trends Biochem Sci* 29, 348–363.

3

Hydrothermal Conditions Are Conducive for the Origin of Life

Volcanological research constitutes an important link in the outcome of
the major worldviews related to the origin of life problem.

A.I. Oparin

Overview and Questions to be Addressed

Alexander Ivanovich Oparin was first to consider the origin of life in strictly scientific terms. Oparin published *The Origin of Life* in 1924, in his native Russian language, and was active in the field for the next 50 years. During my initial field work in the volcanic regions of Kamchatka, organized with Vladimir Kompanichenko, we visited the Institute of Volcanology and Seismology in Petropavlovsk, and I happened to see the above quote painted on a wall near the entrance. Oparin's proposal about how life can begin was intuitive because he had no experimental evidence as a foundation, but as our party rode in helicopters up and down the peninsula from one volcanic site to the next, I began to share his intuition.

The focus of this chapter concerns the properties of water in contact with mineral surfaces heated by volcanism, inspired by what we saw in Kamchatka. Four billion years ago, as the global temperature decreased following the condensation of the ocean, there came a point at which the components required for the origin of life could assemble into systems of encapsulated polymers. Two alternative hydrothermal conditions have been proposed as sites where this could have occurred: salty seawater at submarine hydrothermal vents and freshwater circulating in hydrothermal fields associated with volcanic land masses. To weigh the alternatives, this chapter considers the chemical and physical properties of hydrothermal vents and hydrothermal fields and how each could contribute to the origin of cellular life.

Questions to be addressed:

- What are the chemical and physical properties of hydrothermal vents?
- How do the properties of hydrothermal fields differ from those of vents?
- How are these properties related to the way that organic solutes can undergo physical and chemical interactions related to the origin of life?

Learning from Prebiotic Analogues

Suppose that an organic chemist decides to synthesize a new compound that involves making an ester bond. The chemist is provided with a solution of the two reactants such as acetic acid and ethanol, and then is given a choice: should the reaction be run in an ice bath or instead heated to boiling and refluxed? Usually the choice will be to run the reaction at an elevated temperature. Why is that? The reason has to do with activation energy. All chemical reactions have a measurable energy that must be added to the reactants before they can proceed energetically downhill toward equilibrium, and that energy is most easily added simply by heating the solution. This is why hydrothermal environments seem more conducive than icy conditions for chemical processes related to the origin of life. Furthermore, as noted in Chapter 2, the global temperature of the early Earth at the time of life's origin was much warmer than today, and there may have been negligible ice available, even at the poles. On the other hand, there was abundant heat from volcanism on the early Earth, and the volcanic activity would have involved the hot springs and geysers we refer to as hydrothermal conditions, just as it does today.

Given the fact that most of the Earth's water is in the ocean, it has always seemed reasonable to conclude that life must have begun in salty seawater. But the bulk of the ocean is at equilibrium with little or no free energy available to drive chemical reactions, while the hydrothermal vents discovered in 1977 (Plate 3.1A) are in an obvious state of disequilibrium with available free energy and were immediately recognized to be a potential site for life to begin (Corliss et al., 1981; Baross and Hoffman, 1985). Hydrothermal vents are classified in terms of their temperature, the composition of the vent fluid, and the source of energy that produces them. Black smokers are heated geothermally by underlying magma (Fig. 3.1A). They occur along MORs in deep water (2 km) at pressures (~300 atmospheres) that allow water to remain as a liquid even though its temperature can exceed 400° C. The vent fluid is acidic with pH values ranging from 2 to 3 and there is a steep temperature gradient across the vent walls between the hot vent fluid and the 2° C temperature of seawater at that depth. It is possible to drill into the vent mineral and sample microbial populations. Kashefi and Lovely (2003) observed living microorganisms isolated from black smokers that could be cultured at temperatures up to 121° C, which is the highest temperature known at which microbial life can exist. Black smokers tend to be short lived, with life spans measured in tens to hundreds of years, then reappear elsewhere. The "smoke" of black smokers is composed of metal sulfides that are soluble at the highest vent temperatures but precipitate when they come into contact with cold seawater.

Alkaline vents, sometimes referred to as "white smokers" because they emit a smoke composed of white carbonate minerals, are associated with geological sites undergoing serpentinization (Kelley et al., 2005). When peridotite, the primary mineral component of the Earth's mantle, comes into contact with water, its content of olivine (magnesium-iron silicate) becomes oxidized, with the products being magnetite and serpentine. The reaction generates hydrogen gas and significant amounts of heat, so even though white smokers emit warm water, they are not associated with volcanism. The water coming up through the vents is strongly alkaline, with typical pH values in the range 9–11. When it comes into contact with cold seawater at lower pH (~8) the dissolved minerals precipitate to form calcite and brucite (calcium and magnesium carbonate) which produce the characteristic white alkaline vent structures (Plate 3.1B).

Although the two primary hydrothermal vent structures are products of different sources of energy, they share the fact that they are in constant contact with seawater, which means that condensation reactions relevant to the origin of life have a substantial thermodynamic hurdle to overcome. Life uses condensation reactions to form the ester and peptide bonds that are ubiquitous in linking monomers into the polymers of life—nucleic acids and proteins—but condensation reactions are not spontaneous in aqueous solutions, instead requiring a source of energy.

A Freshwater Alternative to an Origin of Life in Salty Seawater

The possible relationship of hydrothermal vents to the origin of life have been described in multiple publications by Michael Russell, William Martin, and Nick Lane that will be discussed later, so they will not be described in detail here. Instead I will focus on an alternative site associated with volcanic islands emerging from the ocean on the early Earth 4 Gya. We can get some idea of what they were like by visiting volcanic sites on today's Earth. To that end, I have carried out field studies in the Kamchatka peninsula in far eastern Russia, in Iceland, Hawaii, and closer to home in Bumpass Hell on Mount Lassen (Deamer et al., 2006). For a scientist used to working in the sterile environment of a modern laboratory, venturing into a volcanic site on today's Earth is an eye-opening experience. It forces us to question the glib assumption that what happens under laboratory conditions can be translated 4 billion years back in time to the conditions on the prebiotic Earth. These visits have informed laboratory simulations that will be described in Chapter 8.

Because volcanism is central to the scenario being developed, it is worth taking a moment to sketch the major features of volcanoes today, then extrapolate back to the early Earth and Mars 4 billion years ago. Plate 3.2 illustrates the main features of a typical erupting volcano and the aftermath. The eruption is powered by an underlying chamber of molten basalt called magma at a temperature ranging from 700 to 1300° C. As the magma rises through the Earth's crust, at some point it finds a weak region through which the enormous pressures of gases dissolved in the molten basaltic minerals can be released. However, the most spectacular features of volcanoes—eruptions and red-hot lava flows—are not particularly relevant to the narrative being developed here. Instead, we are focusing on the hydrothermal fields that remain after an eruption, and the composition of the gases and minerals that are in a state of disequilibrium with their surroundings and therefore control the composition and chemistry of the water with which they come into contact. It is the physical and chemical properties of the water that are central to the origin of life scenario being presented in this book (Kompanichenko et al., 2015).

The main point is that we can't just think of generic volcanic activity as a driver of life-related chemistry. Instead we must take time and proximity to oceans into account.

The Hawaiian Islands are good examples. The Big Island is the youngest at less than a million years old, and so little time has passed since it formed that rain falling on Kilauea, the active volcano, doesn't accumulate on the surface but instead percolates down through ash, pumice, and lava back into the nearby sea from which it was distilled. There has not been enough time to develop the clays that seal leaks and allow water to circulate in a hydrothermal field. In contrast, volcanoes on Iceland (14–16 million years old) have had enough time for clay, carbonate, and silicate minerals to form, so pools of meteoric water build up and reach a steady state in terms of ionic solutes. Another well-studied example is Yellowstone National Park in Wyoming, which was produced by at least three volcanic eruptions 630,000, 1.3 million, and 2.1 Mya, forming a caldera with dimensions of approximately 54 by 72 km. Yellowstone is a useful prebiotic analogue because it has large numbers of characteristic hot springs, pools, and geysers that give us some idea of how local geochemical processes affect the composition of water and how water interacts with mineral surfaces. In contrast to hydrothermal vents, hydrothermal fields are fluctuating environments with cycles of hydration and dehydration related to evaporation and precipitation, as well as variations in water level caused by geysers and hot spring activity.

Chemistry and Physics of Hydrothermal Fields

An essential feature of the origin of life scenario proposed in this book concerns the way that water can affect the physical and chemical properties of compounds dissolved or dispersed in it. The simplest property is that water is a solvent, so that on the early Earth and Mars the bodies of water had certain dissolved solutes. The global ocean was the most extensive body of water on early Earth, but evidence revealed by three Mars landers tells us that Mars also had shallower seas 4 Gya. One of our assumptions is that the water was salty because it had dissolved various soluble ions from the minerals of the crust during the violent stirring that must have occurred when the oceans condensed from the atmosphere. The major source of freshwater would have been precipitation that fell on volcanic land masses emerging above sea level on both planets.

Figure 3.1 shows the concentrations of the main ionic solutes in terrestrial seawater on a linear and logarithmic scale, and for comparison the ionic solutes in the water of a typical hydrothermal field; Table 3.1 has exact values. Sodium chloride is by far the highest concentration in seawater at ~0.5 M, followed by 53 mM magnesium, 10 mM calcium, and 10 mM potassium. In contrast, the ionic solutes in hydrothermal water are much more dilute, ranging from <1 to ~17 mM.

Why is the salt content an important point? There are three reasons. First, in today's ocean, sodium chloride is at the highest concentration. If salt concentration was not balanced across its membrane, a primitive cell would collapse because of the large osmotic pressure that would develop. The self-assembly of lipids into closed vesicles is a spontaneous process, but it is significantly impeded in high salt concentrations. In fact, all life today has energy-dependent mechanisms to regulate volume, and there are no cells that function with high concentrations of intracellular sodium chloride. Instead sodium is actively pumped out of cells in exchange for potassium which is maintained at an internal concentration of approximately 0.1 M. This fact led to speculations that life

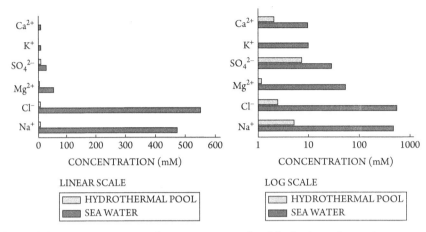

Figure 3.1. Ionic composition of seawater compared with hydrothermal water in a hot spring on the flank of Mount Mutnovsky in Kamchatka, log and linear scales. Hydrothermal data from Kompanichenko et al. 2016.

Table 3.1 **Ionic Composition of Seawater**

Seawater pH = 8.1	
Na	469 mM
Cl	546 mM
Ca	10 mM
Mg	53 mM
K	10 mM
HCO_3	2 mM
SO_4	28 mM
SiO_2	0
Cationic equivalents	605 mM
Anionic equivalents	604 mM

Source: Millero et al., 2008.

began not in the high concentrations of sodium chloride found in seawater but instead in hydrothermal fields in which potassium chloride was among the main ionic species present (Mulkidjanian et al., 2012).

Second, the divalent cation content of seawater would be problematic for assembly of simple amphiphilic molecules like fatty acids into closed membranous compartments. Calcium and magnesium concentrations in seawater are 10 and 54 mM respectively, and calcium, in particular, strongly interacts with the anionic carboxylate and phosphate groups of lipids, which results in insoluble calcium soaps that precipitate as aggregates rather than forming membranes. (Anyone who has tried to wash their hands in seawater or other sources of hard water has experienced this effect.) All cells today actively pump calcium ions outward, and a typical cytoplasmic concentration is maintained in the range of 10–100 nM.

Third, the inhibitory effects of calcium are not limited to lipids, because at the alkaline pH of seawater, phosphate precipitates to form a virtually insoluble calcium phosphate mineral called apatite. Many important organic compounds that contain phosphate are also precipitated as calcium complexes. However, the solubility problem might be easily resolved, as discussed in the next section.

Apatite Mineral Is a Special Case

We have focused on calcium and magnesium minerals because they are the primary divalent cations in seawater and are also intimately related to life processes. As a general rule, minerals composed of a divalent ion like calcium are much less soluble than those composed of a divalent cation and a monovalent anion. For instance, calcium sulfate composes the relatively insoluble mineral gypsum while calcium chloride is soluble in water. The reason it is important to understand mineral solubility is that phosphate is an essential component of all life today. It is a linking molecule in the polymer structure of nucleic acids, and the phosphate anhydride bond in adenosine triphosphate (ATP) is the primary energy currency for metabolism. Enzyme-catalyzed ATP hydrolysis also drives the transmembrane ion transport processes required for maintaining ion concentration gradients between the intracellular cytoplasm and the external environment, as well as the contraction of actin and myosin in muscle fibers.

Despite the central role of phosphate in biology, it is not at all clear how it first became incorporated in the primitive metabolism of early cellular life. Soluble phosphate is rare in today's biosphere, and it has been assumed that it would also be rare in the prebiotic environment, because the most abundant mineral form of phosphate is apatite, in which calcium and phosphate form microscopic crystals that have a very low solubility at neutral and alkaline pH ranges. This relative insolubility is made clear by the fact that apatite is the primary mineral of tooth enamel which is stable for many years under ordinary circumstances. The apparent lack of soluble phosphate in the prebiotic environment is referred to as the phosphate problem.

On the other hand, it is common knowledge that bacterial films adhering to tooth enamel can generate an acidic condition sufficient to dissolve the enamel and produce a cavity. The main acid is lactic acid, a byproduct of bacterial metabolism, and the measured acidity under a bacterial film is pH 4.5. From this observation, it is equally clear that it only requires a moderate acidity to dissolve the apatite crystals in enamel, thereby releasing calcium and phosphate into solution.

Is it possible that the phosphate problem can be resolved simply by assuming that life began in an acidic medium? Let's consider what could lead to acidic bodies of water on the early Earth. The most obvious is the acidic pH produced by carbon dioxide which forms carbonic acid when dissolved in water. For instance, if pure water at pH 7.0 is exposed to air in which carbon dioxide is 0.4×10^{-3} atmospheres, the pH will slowly decrease over time to approximately 5.65. If water is exposed to 1.0 atmosphere of carbon dioxide the pH falls to 3.92. It follows that carbon dioxide alone can produce an acidity that may keep phosphate in solution. An experiment showing this is illustrated in Figure 3.2. A solution of 1.0 M dibasic sodium phosphate was added to 100 mL of seawater to a final concentration of 10 mM. Note that the seawater, clear before the phosphate addition (Fig. 3.2A), immediately becomes turbid due to the precipitation of apatite after phosphate is added (Fig. 3.2B). Figure 3.2C shows the same flask after being flushed with carbon dioxide for a few minutes. The apatite has entirely dissolved.

This simple experiment shows that small bodies of freshwater in hydrothermal fields associated with volcanic land masses have the potential to contain soluble phosphate at reasonable concentrations if they are in contact with mineral apatite. However, it is unlikely that carbon dioxide was equivalent to 1 atmosphere 4 billion years ago. Are there other sources of acidity? Sulfur dioxide (SO_2) is a common constituent of volcanic emissions, as well as hydrogen chloride (HCl). When SO_2 reacts with water, it produces sulfurous acid, which transforms into sulfuric acid upon further oxidation. The acidity of the hydrothermal water of Mount Mutnovsky in Kamchatka and Bumpass Hell on Mount Lassen is ~pH 3. Similar pH ranges have been reported for hydrothermal ponds in Yellowstone National Park, as will be described in the next section. The pH of hydrothermal water becomes much more acidic when the water undergoes evaporation. For instance, on the small volcanic White Island off the coast of New Zealand, the pH of the small pond in the island's center is pH 0, equivalent to 1 M sulfuric acid! Despite this low pH, abundant diatoms and motile bacteria thrive in in the pond water.

From this, it should be clear that the aqueous environment most conducive for the origin of cellular life on the early Earth would have had dilute concentrations of salts like sodium and potassium chloride and divalent cation concentrations in the mM range. It is amusing to point out that biochemists have taken this lesson to heart and use deionized water for their experiments, typically buffered at neutral pH ranges. If a salt is required, it would be potassium chloride at a concentration of 0.1 M, and 5 mM MgCl2

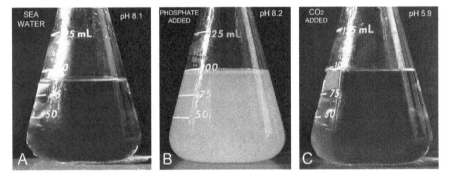

Figure 3.2. When a solution of 1.0 M dibasic sodium phosphate was added to 100 mL of filtered sea water (A) to a final concentration of 10 mM, the seawater becomes turbid (B). This turbidity is caused by microscopic crystals of calcium phosphate minerals that form at alkaline pH ranges. When carbon dioxide is introduced to the seawater (C), the pH becomes more acidic (pH 5.9) and the mineral crystals dissolve.

required by certain enzymes. Virtually no one would collect a liter of seawater to be used in biochemical experiments, because the salt and divalent cation concentrations would inhibit the functions of many essential enzymes.

Ionic Solutes of Hydrothermal Fields

Table 3.2 shows the concentrations of ionic solutes in two Yellowstone hot springs (Rowe et al., 1965). The water in Mammoth Hot Springs is near neutral pH and is characterized by its content of bicarbonate, sulfate, and calcium ions. As a result, when the hot spring water comes into contact with the atmosphere and begins to evaporate, calcium carbonate and calcium sulfate precipitate to produce the massive mineral deposit that gave the hot spring its name.

In contrast, the Norris Geyser Basin water is more acidic (pH 3–4), has a higher temperature range (80–90° C), and has more sodium and silicate in solution along with the sulfate, with only traces of divalent cations.

The reason this is important is that in later chapters we will propose that the energy required to cause evaporation can be transduced into chemical potential necessary to drive polymerization. Part of the energy emerges from the decrease in entropy resulting from concentrating potential reactants and the order imposed on those reactants as they become crowded in nearly anhydrous films. However, laboratory simulations generally use pure water and reagents, so it important to remember that the organic solutes might be concentrated, but so are the ionic salts that are present. In our research, we have taken both of these points into consideration.

Setting the Stage

Hydrothermal black smokers are generated by magma rising to the Earth's crust along MORs where the Earth's crust is significantly thinner than the continental crust. However, these are relatively short lived, so attention has shifted to alkaline hydrothermal vents generated by serpentinization. Alkaline vents composed of carbonate mineral exist for thousands of years and have substantial free energy available in sustained far-from-equilibrium conditions. These have been proposed as sites at which chemical reactions could initiate primitive metabolism involving the reduction of carbon dioxide by dissolved hydrogen.

An alternative site is associated with volcanoes emerging through the global ocean. The existence of early terrestrial volcanoes is supported by the fact that volcanism was also occurring on Mars over 3 billion years ago. Precipitation on volcanic landmasses would produce hydrothermal fields characterized by freshwater geysers, hot springs, and clay-lined pools that undergo cycles of hydration and evaporation.

A recent discovery led to a significant conclusion that is consistent with the narrative of this book. As described in Chapter 1, the earliest biosignatures of microbial life are the stromatolite fossils found in the Dresser formation of the Pilbara region of western Australia. Because stromatolites today are associated with shallow saline environments such as the Hamelin Pool in Shark Bay, Australia, it has been assumed that the fossil

Table 3.2 **Ionic Composition of Representative Yellowstone Hot Springs**

Mammoth Hot Springs T = 71.5 +/– 2°C (N =7)			*Mean*
pH	6.5 +/– 0.3	(N = 6)	6.5
Na	131 +/– 7 ppm	(N =8)	5.7 mM
Cl	166 +/– 3 ppm	(N = 9)	4.6 mM
Ca	305 +/– 60 ppm	(N = 6)	7 mM
Mg	68 +/– 5.7 ppm	(N = 7)	3 mM
K	55 +/– 8 ppm	(N = 7)	1.4 mM
HCO_3	725 +/– 135 ppm	(N = 5)	12 mM
SO_4	510 +/– 38 ppm	(N = 7)	5 mM
SiO_2	53 +/–3.8 ppm	(N = 8)	0.9 mM
Cationic equivalents	27 mM		
Anionic equivalents	26.6 mM		
Norris geyser basin T = 86 +/–6°C (N = 20)			
pH	4.1 +/– 0.7	(N = 14)	4.1
Na	299 +/– 130 ppm	(N = 18)	13 mM
Cl	611 +/– 171 ppm	(N = 10)	17 mM
Ca	5.9 +/– 3.8 ppm	(N = 16)	0.120 mM
Mg	0.6 +/– 0.3 ppm	(N = 16)	0.026 mM
K	53 +/– 26 ppm	(N = 38)	1.4 mM
H_2CO_3	~0		
SO_4	131 +/– 68 ppm	(N = 4)	1.3 mM
SiO_2	450 +/– 123 ppm	(N = 9)	<u>7.5 mM</u>
Cationic equivalents	14.5 mM		
Anionic equivalents	18.3 mM		

Note: Concentrations are shown as parts per million (ppm) and calculated as millimolar or micromolar equivalents.

Source: Rowe et al., 1965.

stromatolites developed in a marine environment. However, Djokic et al. (2017) reported that geyserite, a silicate mineral exclusively associated with freshwater hydrothermal conditions, is present in conjunction with the stromatolite fossils (Fig. 3.3 and Plate 3.3) and concluded that:

> The Dresser Formation terrestrial hot spring facies include geyserite, siliceous sinter terracettes and the mineralized remnants of hot spring pools. These findings extend the geological record of inhabited terrestrial hot springs by

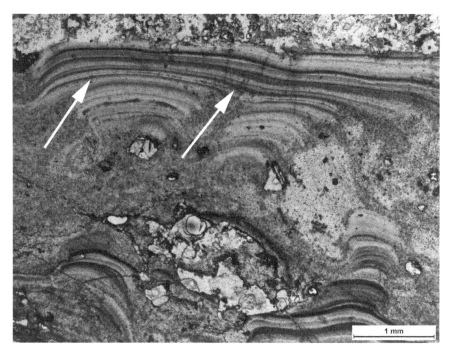

Figure 3.3. Geyserite in the Dresser formation is consistent with hydrothermal field conditions rather that a shallow marine setting for the Pilbara Craton. The thin multiple layers (arrows) are characteristic of silicate minerals deposited by cyclic eruptions of geysers. Photograph by Tara Djokic (Djokic et al., 2017).

~3 billion years, the occurrence of an exposed land surface by up to ~130 million years and evidence of life on land by ~580 million years. This result is significant in that it further constrains our understanding of the evolution of early life on Earth, as well as offers astrobiological implications in the search for potential fossil life on Mars. The Dresser Formation shares a similar age to older portions of the martian crust and provides the closest comparison to geological processes likely occurring on Mars at that time.

Summary

Hydrothermal fields and alkaline hydrothermal vents are both attractive as sites that would be conducive to the origin of life. We will conclude this chapter by summarizing the relative merits and limitations of both (Deamer and Georgiou, 2015).

Favorable Properties of Alkaline Hydrothermal Vents

- A source of chemical energy is available from solutes and minerals at different redox states. Energy may also be available in the substantial pH gradients that form when alkaline seawater in the porous minerals at high pH comes into contact with lower pH bulk seawater.

- The estimated life of alkaline vents can be thousands of years, providing a continuing supply of chemical energy at life-compatible temperature ranges (50–90°C).
- Vent minerals are a source of transition metals with potential catalytic activity when incorporated into peptides.
- Small amounts of formic acid were detected in simulations of hydrothermal vents, suggesting that dissolved carbon dioxide was reduced to formic acid.
- A mechanism for concentrating dilute solutes has been demonstrated in a laboratory simulation of alkaline vents.

Limitations of Alkaline Hydrothermal Vents

- Self-assembly of amphiphiles into membranes is inhibited by high salt concentration and divalent cations present in alkaline vent fluids.
- Condensation reactions leading to polymerization require activated substrates in an aqueous medium like seawater. Neither activated substrates nor polymerization processes have been demonstrated experimentally in hydrothermal vent conditions.
- The proposed chemiosmotic potential of pH gradients across mineral membranes of vents has not yet been demonstrated.
- Cycling does not occur in hydrothermal vent conditions.
- Photosynthesis could not develop in vent conditions.

Favorable Properties of Hydrothermal Fields

- A source of chemical energy is available from reduced water activity during dehydration.
- Concentration of dilute solutes on mineral surfaces occurs naturally upon drying.
- Self-assembly of amphiphilic compounds readily occurs in low ionic strength freshwater.
- Condensation reactions and polymerization have been demonstrated in laboratory simulations.
- Cycles of hydration and dehydration drive increased complexity as products accumulate in closed systems of hydrothermal pools.
- Light energy is abundant, so photosynthesis can develop.
- Phosphate minerals are soluble in acidic hydrothermal water.

Limitations of Hydrothermal Fields

- Low pH ranges may inhibit certain reactions.
- Extensive clay deposits adsorb organic solutes and make them unavailable as potential reactants in solution.
- Exposure to the ultraviolet component of sunlight can degrade certain molecules.

References

Corliss JB, Baross JA, Hoffman SE (1981) An hypothesis concerning the relationship between submarine hot springs and the origin of life on Earth. *Oceanol Acta* 4, 59–69.

Baross JA, Hoffman SE (1985) Submarine hydrothermal vents and associated gradient environments as sites for the origin and evolution of life. *Orig Life Evol Biospheres* 15, 327–345.

Deamer D, Singaram S, Rajamani S, Kompanichenko V, Guggenheim S (2006) Self-assembly processes in the prebiotic environment. *Phil Trans R Soc Lond B* 361, 1809–1818.

Deamer DW, Georgio CD (2015) Hydrothermal conditions and the origin of cellular life. *Astrobiology* 15, 1091–1095.

Djokic T, Van Kranendonk MJ, Campbell KA, Walter MR, Ward CR (2017) Earliest signs of life on land preserved in ca. 3.5 Gya hot spring deposits. *Nat Commun* 8, 15263. doi: 10.1038/ncomms15263.

Kashefi K, Lovley DR (2003) Extending the upper temperature limit for life. *Science* 301, 934.

Kelley DS et al. (2005) A serpentinite-hosted ecosystem: The Lost City hydrothermal field. *Science* 307, 1428–1434.

Kompanichenko VN, Poturay VA, Shlufman KV (2015) Hydrothermal systems of Kamchatka are models of the prebiotic environment. *Orig Life Evol Biosph* 45, 93–103. doi: 10.1007/s11084-015-9429-2.

Mulkidjanian AY, Bychkov AY, Dibrova DV, Galperin MY, Koonin EV (2012) Origin of first cells at terrestrial, anoxic geothermal fields. *Proc Natl Acad Sci USA* 109, 21–E830.

Rowe JJ, Fournier RO, Morey GW (1965) Chemical analysis of thermal waters in Yellowstone National Park, Wyoming, 1960–65. *Geological Survey Bulletin* 1303.

4

Sources of Organic Compounds Required for Primitive Life

All of the rocky and metallic material we stand on, the iron in our blood,
the calcium in our teeth, the carbon in our genes were produced billions
of years ago in the interior of a red giant star. We are made of star-stuff.

Carl Sagan

Overview and Questions to be Addressed

Carbon compounds on the early Earth were not the simple mixture previously
referred to as a "prebiotic soup." Instead, there was a continuing input of organic
material synthesized by geochemical and photochemical reactions in the volcanic
crust and atmosphere; organic compounds were also being delivered during late
accretion by the infall of interplanetary dust particles (IDPs), impacting comets,
and asteroid-sized bodies. Compounds from both sources (terrestrial and not) then
underwent chemical processing by volcanism, photochemistry, and mineral-driven
oxidation–reduction reactions. Some of these processes were synthetic reactions
that led to increasing complexity, but this was balanced by other processes such as
hydrolysis and pyrolysis that degraded organic material into simpler compounds
or tar-like polymers. Because the atmosphere contained no molecular oxygen, the
organic compounds that formed were relatively stable as a dilute solution in the
global ocean, but were also dissolved in freshwater hydrothermal pools in contact
with mineral surfaces of volcanic land masses. In either case, a process was required
by which the organic compounds could become sufficiently concentrated to undergo chemical reactions.

Questions to be addressed:

- What are plausible sources of organic compounds?
- What is their composition and abundance?
- How would organic material be chemically processed on the early Earth?
- How can dilute organic solutes become sufficiently concentrated to undergo chem-
  ical reactions?

Sources of Organic Carbon Compounds

Chapter 1 described how virtually all of the carbon now circulating in the biosphere as organic and inorganic compounds was delivered during accretion of planetesimals as the Earth formed, and it is reasonable to assume that Mars had a similar addition of carbon compounds and water after it cooled from primary accretion. On the Earth, organic substances delivered during primary accretion would have been destroyed by the heat of impacts and the moon-forming event, so the carbon compounds necessary for the origin of life were necessarily added after the Earth had cooled sufficiently for a global ocean to appear. Mars did not undergo a moon-forming collision, so an interesting possibility is that conditions conducive to the origin of life were present on Mars long before the Earth cooled sufficiently for a global ocean to condense. On both planets, two sources of organic compounds, not mutually exclusive, are: delivery as extraterrestrial infall during several hundred million years of late accretion (Chyba and Sagan, 1992) and geochemical reactions that could synthesize organic compounds (Bada and Lazcano, 2003). It is still uncertain what the relative fraction or composition of these two sources was, but the composition of comets and carbonaceous meteorites can be used as a guide to the compounds likely to be available for the origin of life.

There are several important points to be made about the history of these compounds and their solutions in bodies of water. We will begin by summarizing known sources of organic compounds present in the early solar system and then evaluate which could have contributed to the suite of organics available for the origin of life on the Earth and Mars. (See Ehrenfreund and Cami, 2010, for a thoughtful review and summary of this process.)

Stellar Nucleosynthesis and the Death of Stars

The first stars that appeared in the universe after the Big Bang used hydrogen fusion as an energy source, just as stars do today. The primary product of fusion reactions is helium, but it is now known that toward the end of their lifetimes stars ultimately exhaust their supply of hydrogen and become so hot that a series of secondary fusion reactions begin. These produce heavier atoms up to iron, including the biogenic elements of carbon, oxygen, nitrogen, phosphorus, and sulfur (CONPS) that are essential for life. An important point is that elemental silicon is also synthesized, later to become the silicate minerals of interstellar dust that finally accrete into planets. At some point ordinary stars exhaust this fusion energy source as well. They first expand into a red giant phase, then collapse into an extremely dense white dwarf. During the collapse, much of their mass is ejected into space to form the interstellar dust that gathers into vast molecular clouds throughout galaxies (Plate 4.1)

This history is explained here because virtually all stars and planets today assemble within molecular clouds composed of the dusty ashes of stars that existed billions of years ago. Observations of molecular clouds such as the Orion nebula by the Hubble telescope have revealed newly formed stars with obvious disks of dust around them and even hints of large planets within the disks. Based on this evidence, Plate 4.2 illustrates the appearance of the early solar system.

Microwave emission spectra and infrared absorption spectra can provide information about the composition of molecular species in outer space. The astonishing observation is that carbon, nitrogen, and oxygen in molecular clouds are present as hundreds of organic compounds (Garrod et al., 2008). A list of biologically relevant compounds confirmed by radio astronomy is provided here. Many other compounds have also been detected but can only exist at the extremely low temperatures of molecular clouds.

CO carbon monoxide
CO_2 carbon dioxide
HCN hydrogen cyanide
NH_3 ammonia
H_2O water
C_2H_2 acetylene
HCHO formaldehyde
CH_4 methane
CH_3OH methanol
HCOOH formic acid
NH_2CN cyanamide
$HCONH_2$ formamide
CH_3COOH acetic acid
$(CH_3)_2CO$ acetone
CH_3CONH_2 acetamide
$(CH_2OH)_2$ ethylene glycol
CH_2NH_2COOH glycine

How could these organic compounds be synthesized? One possibility is that the interstellar dust particles in molecular clouds are coated with thin layers of ice mixed with simple molecules like carbon monoxide, carbon dioxide, methanol, and ammonia. Laboratory simulations of grain photochemistry at very low temperatures have confirmed that synthetic reactions can be driven when ultraviolet photons impinge on ice layers composed of these compounds (Allamandola et al., 1988). The photochemical products suggest that complex molecules first accumulated on the surfaces of dust particles and then into the planetesimals that gave rise to the early solar system. Asteroids and comets are examples of planetesimals, and both contain significant amounts of organic compounds that have been preserved since the origin of the solar system ~4.6 Gya.

Organic Compounds in Comets

The usual definition of a comet applies to their visual appearance when seen from the Earth, mostly as a long glowing tail caused by the release of dust and gas when a small icy nucleus is heated by the sun. In the discussion to follow, it is convenient to use the term comet with the understanding that we are referring specifically to the nucleus. The early solar system was composed of a disk of dust slowly revolving around a central mass they would become the sun. Within the disk, eddys under the influence of gravity grew into vast numbers of kilometer-sized planetesimals composed of silicate minerals, ice, iron and the organic compounds on the dust surfaces. Most of the planetesimals accreted into planets, but the comets in the Kuiper belt and Oort cloud in the outer solar system, as well as the asteroids in orbit between Mars and Jupiter, are remnants that did not get caught up in planet formation. Comets are defined by their substantial content of water ice, typically more than half their mass (Plate 4.3). A smaller fraction is mineral dust, presumably the original micron-sized silicate mineral particles that accumulated in molecular clouds. It has also been estimated that as much as a tenth of the mass of a comet is organic matter that was originally bound to the surfaces of the dust. These include reactive species such as formaldehyde and HCN, along with polycyclic aromatic material (Mumma and Charnley, 2011).

A simple calculation gives a perspective on the amount of organic carbon that could have been delivered intact to take part in the initial chemical evolution toward life. First, we can ask how much organic carbon might have been sufficient to initiate the chemical and physical processes leading to the origin of life. Obviously, the amount in today's biosphere is sufficient for all life today and has been estimated to be approximately 6 $\times$ 10^{15} kg. That sounds like a lot, but in fact if all of the biomass could be spread evenly over the Earth's surface area it would form a layer ~10 mm thick. Furthermore, if the biomass could somehow be broken down into monomers like amino acids and the glucose content of cellulose and then dissolved in the ocean, the solution would be very dilute, about 5 micromolar. At this concentration there are 10 million water molecules for every monomer molecule. Because molecules need to collide in order to react, intermolecular collisions and reactions would be rare events in such dilute solutions, so the origin of life must have required much more concentrated reactants. For instance, the biochemical metabolites in living cells are in the millimolar range, a thousand times more concentrated. This is why it is necessary to establish plausible mechanisms by which organic solutes in the prebiotic environment can be concentrated in order for life to begin.

Now, we can estimate the amount of water and organic material that could have been supplied by comets during late accretion. The mass of a typical comet is 10^{14} kg, a number easily calculated from its dimensions and density. The mass of the ocean is 10^{21} kg, so the accretion of 10 million comets would be required to provide all of the Earth's seawater. However, as discussed in Chapter 1, the source of ~90% of the ocean's water was outgassing from the Earth's interior, leaving an estimated ~10% to be delivered by a million comets. Comets are ~10% organic carbon, so a typical comet contains 10^{13} kg of mixed organic compounds. Taking 6 $\times$ 10^{15} kg as the amount of organic material in the biosphere, it would require just 600 comets out of the million total to deliver the organic carbon now circulating in the biosphere!

From calculations like this, it seems possible that comets are an obvious source of organic compounds as suggested by Oro (1961) and later promoted by Delsemme (2000), but this arithmetic assumes that all of the organic material in a comet can survive the immense energies released when it impacts a global ocean. An example of such energy release is provided by the 10 km object that struck the Earth 66 million years ago near what is now the Yucatan peninsula, causing the most recent major extinction event in Earth's history. Even though this was a stony asteroid rather than a comet, there is little doubt that most (but not all) of the organic material in a comet impacting the ocean would be largely reduced to water vapor and simple carbon compounds such as CO_2. One might imagine that landing in the ocean would somehow cushion the impact, but the ocean is only 5 km deep and a typical comet is 5 to 15 km in diameter, so an incoming comet traveling at 20 km per second would penetrate the ocean thickness in less than a second and then vaporize when it encountered the rocky ocean floor. This is well outside ordinary human experience, but a simple calculation can give a perspective. The kinetic energy of an object is given by the formula $E = (mV^2)/2$. If the mass of a comet is 10^{14} kg and it arrives at a velocity of 2×10^4 m/sec, the energy released upon impact is 4×10^{22} Joules. The largest hydrogen bomb ever tested produced 6×10^{15} joules of energy, so the energy of an impacting comet, all released in less than a second, would be like 7 million hydrogen bombs all exploding simultaneously. The ice of the comet would vaporize in a white-hot flash, and any organic material would be degraded to CO_2 and water vapor. Despite this, tests have shown that a small amount of organic matter would survive, blown away by the explosion before being subjected to the intense heat.

The tentative conclusion is that comets might supply a small fraction of the organic compounds required for life to begin, but as will be discussed later in this chapter, a much more substantial source is available.

Carbonaceous Meteorites

Asteroids undergo continuing collisions that produce smaller fragments, some of which migrate inward toward the sun. A few of these intersect the orbits of the inner planets and fall to the planetary surface as meteorites. This means that most meteorites are samples of the surfaces of asteroid parent bodies, but the asteroids have experienced significantly different histories in the early solar system. Larger asteroids ranging from a few to several hundred kilometers in size have undergone differentiation due to the decay of radioactive elements like aluminum 26 that heated their interiors. The temperatures increased to the point that metallic iron and nickel melted to form a core, which is the source of iron-nickel meteorites common in museum collections. During the heating process some of the original organic material was altered by pyrolysis, while other compounds were subjected to a natural version of steam distillation as water melted and vaporized. The liquid water and steam transported the organic compounds toward the surface where they accumulated. Wing and Bada (1991) referred to this process as geochromatography and speculated that it could explain the differential composition of organic material observed in some meteorites.

Smaller asteroids ranging from meters to a few hundred meters in size do not have enough mass to have been heated and are relatively undifferentiated. The meteorites called carbonaceous chondrites are samples of their surfaces and provide the most significant information about the organic compounds present in the early solar system prior to and during planet formation. The Murchison meteorite exploded over the town of Murchison, Australia, in September 1969, and is one of the best studied carbonaceous

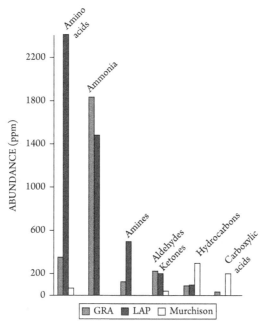

Figure 4.1. Comparison of Murchison organic composition with that of two meteorites collected in Antarctica, labeled GRA and LAP. Adapted from Pizzarello and Shock 2010.

chondrites, along with several others collected in the ice fields of Antarctica. Figure 4.1 and Plate 4.4 summarize the main organic compounds present in the Murchison meteorite and two other carbonaceous meteorites. As might be expected, there is a surprising degree of variability in composition that can be accounted for by the different histories of the parent bodies (Pizzarello et al., 2011).

We will focus on the Murchison composition as a guide to the kinds of organic compounds that could have been delivered to the prebiotic Earth's surface. The main mass of the Murchison meteorite is present as microscopic grains of silicate minerals such as olivine and pyroxene. The organic material is mostly a kerogen-like polymer that composes approximately 1.5% of the total (Plate 4.4). About one part in a thousand of the mass is soluble in water, the most familiar being ~70 amino acids dominated by glycine and alanine, with eight others also present in biological proteins. This composition is similar to the amino acids synthesized in the original Miller experiment, and later studies suggested that the amino acids are products of the Strecker synthesis in which formaldehyde reacts with HCN. Also soluble in water is a series of monocarboxylic acids ranging from formic and acetic acid up to the 12-carbon dodecanoic acid. As will be described in Chapter 5, the longer-chain monocarboxylic acids are able to assemble into membranous structures essential for the origin of cellular life.

Significantly, heterocyclic compounds are also present, including nucleobases that were produced by nonbiological chemical processing (Martins et al., 2008). It is important to understand that the organic composition of the Murchison meteorite is not representative of all carbonaceous chondrites. For instance, two Antarctic meteorites have relatively little content of carboxylic acids and hydrocarbons but are rich in ammonia, amino acids, and amines (Fig. 4.1, adapted from Pizzarello and Shock, 2010).

Because carbonaceous meteorites represent a tiny fraction of all the material that would have contributed to late accretion, we cannot assume that their components were a primary source of prebiotic organic material. Instead, their composition confirms that mixtures of organic compounds relevant to life can be synthesized by nonbiological chemical reactions and would be available on the prebiotic Earth, either produced by geochemical and atmospheric reactions or delivered during late accretion by the source described in the next section.

Interplanetary Dust Particles

As illustrated in Plate 4.2, our solar system began as a molecular cloud in which dust particles accumulated into icy planetesimals orbiting around the protosun. These in turn underwent gravitational accretion to form ever larger planetary bodies ranging in size from the inner rocky planets to the gas giants of the outer solar system. Despite the remarkable mixture of organic compounds present in comets and meteorites, most of the organic carbon was probably delivered to the early Earth in the form of IDPs. This was first pointed out by Anders (1989) and later confirmed by Chyba and Sagan (1992). It has been estimated that even today approximately 30,000 tons of extraterrestrial material enters the Earth's atmosphere every year as IDPs (Love and Brownlee, 1991) and micrometeorites. Anyone who wishes to confirm that extraterrestrial infall is still occurring today can do so by running a strong magnet through the runoff from the roof of a house after a rainstorm. The magnet collects spherical iron particles up to a millimeter in diameter produced from metallic micrometeorites that melted upon entering the upper atmosphere.

In the cold vacuum of space, the IDPs are coated with a thin layer of ice, with organic compounds produced by photochemistry in the original molecular cloud (Matrajt et al., 2011). The IDPs enter the atmosphere at the same velocity as larger meteorites, 11 to 75 km per second. Most meteors we see during showers such as the Leonids and Perseids are produced by particles the size of sand grains that are released from comets. When the Earth passes through a previous orbit of a comet, the particles enter the upper atmosphere, lasting for a second or two as they vaporize in the white heat of friction. However, IDPs are much smaller, just several micrometers in diameter (the size of a bacterial cell) and are slowed by atmospheric friction in a few milliseconds. In that brief time, the particle goes from outer-space temperature and passes through a temperature at which the surface ice sublimes in a vacuum nearly that of outer space, depositing water vapor and intact organic compounds in the upper atmosphere. We know that the final temperature is not hot enough to destroy the organic coat on the silicate minerals because the dust particles collected in the upper atmosphere show little evidence of melting (Glavin and Bada, 2004).

Even though ice and volatile organics would flash evaporate when IDPs enter the atmosphere, more robust organic material does adhere to the surface or is embedded in the mineral matrix (Clemett et al., 1993). Many interplanetary dust particles have been captured by researchers who fly aircraft with collecting plates at high altitudes in the upper atmosphere (Messenger et al., 2015) . Analysis of the IDPs requires specialized imaging methods in which an electron beam scans the particles and dislodges small quantities of surface atoms and molecules which are then analyzed by mass spectrometry. Some of the particles still retain organic material on their surfaces. Plate 4.5 shows a scanning microscope image of two such dust particles with organic material marked in green and mineral components in red.

Photochemical Synthesis on Simulated Precometary Grains

Where does the organic content of IDP come from? In pioneering studies, Mayo Greenberg (2002) investigated photochemical reactions that occur when ultraviolet light is absorbed by thin layers of ice mixed with simple carbon compounds and ammonia. The results inspired Lou Allamandola and his coworkers at NASA Ames in Mountain View California to develop a simulation of what is referred to as precometary dust (Allamandola et al., 1988). The goal was to establish the composition of organic compounds that can be synthesized when ultraviolet photons interact with a mixed ice on the surface of a dust particle in a molecular cloud. This involved a metal surface cooled to 15K in an evacuated chamber with a port through which a gas mixture could be introduced. A typical mixture included CH_3OH, NH_3, CO, and water vapor which coated the metal surface with a thin film of mixed ice. The film was illuminated by short wavelength UV light (121 and 100 nm) to simulate the kinds of photons that would impinge on similar icy films on interstellar dust particles. Over a period of days, the film could be seen to take on a yellow color; when the run was over a complex organic mixture remained on the metal surface. It was obvious that the energy of the photons had been captured in the ice and activated a series of photochemical reactions in which the methanol and ammonia molecules had polymerized into more complex structures (Dworkin et al., 2001).

A reasonable conclusion is that there may have been a continuing addition of intact organics to the atmosphere of the early Earth when IDPs entered the atmosphere and were heated to a temperature that flash-evaporated intact volatile compounds. The particles themselves with the remaining organic load would also have survived. Most of the organics would eventually reach the ocean as rainfall and disappear into a very dilute solution. However, an important point is that any portion falling onto volcanic land masses would accumulate on mineral surfaces and be washed into hydrothermal water where it would become increasingly concentrated. This concept was tested by Pearce et al. (2017) who performed a numerical computation of the fate of adenine delivered by an impacting carbonaceous meteorite. The surprising result was that this source of organic material would end approximately 4.17 Gya because of a rapid decrease in accretion of meteoritic infall. Furthermore, if ribonucleic acid (RNA) was synthesized from the adenine, this must occur very quickly because of seepage from the pool and photochemical damage caused by UV light. Thus, volcanoes not only are potential sources of organic compounds, as described in the next section, but in the aftermath of eruptions the volcanic land masses also serve as a site on which organic material can accumulate, in contrast to a global ocean in which organic solutes from infall would become an extremely dilute solution.

Terrestrial Synthesis of Organic Compounds

A possible alternative to extraterrestrial infall as a source of organics is synthesis of a variety of organic compounds in the early Earth's atmosphere and by volcanic geochemistry. The four types of organic compounds that compose all life as monomers, structural subunits, and metabolic intermediates are amino acids, nucleobases, carbohydrates, and amphiphilic compounds such as long-chain monocarboxylic acids. For life to begin, there must have been a way for all of these compounds to accumulate in the prebiotic environment and then somehow assemble into living systems. There is no doubt that they can be synthesized by nonbiological reactions because they are present in carbonaceous meteorites. Furthermore, it is equally certain that they were being delivered to the early Earth because they are still being delivered today. However, as soon as they arrive, they would begin to undergo chemical and physical processing at varying rates, including decomposition into simpler compounds that could not play a role in life's origins. Carbohydrates are most labile to decomposition, followed by amino acids and nucleobases, while hydrocarbon chains are the most stable.

One way to understand the relative stability of hydrocarbons is to consider sources of fossil fuels such as petroleum. The hydrocarbon chains derived from the lipid content of microbial life have survived high temperatures and pressures for several hundred million years, a tenth of the age of the Earth, but there is no trace of the amino acids, carbohydrates, and nucleobases that composed a substantial proportion of the original biological material. This suggests that some fraction of the more labile compounds incorporated into the earliest life were not simply delivered but were being continuously synthesized by atmospheric photochemistry and geochemistry at the Earth's surface. We will discuss each of the four compounds in turn but will first touch on a question that is central to the origin of life but has received relatively little attention.

Sources of Reactive Nitrogen and Amino Acids

Nitrogen is an essential component of proteins and nucleic acids, the two main polymers of life. We live in an atmosphere that is 80% nitrogen, and the atmosphere of the early Earth probably contained an even higher percentage. The reason nitrogen gas is so dominant is that it is among the most chemically inert gas molecules known. Even today, life depends on nitrogen-fixing metabolic reactions of specialized bacteria in a symbiotic relation with higher plant life. So, the question is: How was nitrogen incorporated into the organic chemicals required for the origin of life?

One possibility is that nitrogen was present in the highly reactive compound HCN. This was first pointed out by Stanley Miller, who became a graduate student at the University of Chicago in 1951 and decided to work with Harold Urey for his doctoral research. Urey was interested in the chemical reactions that might occur in mixtures

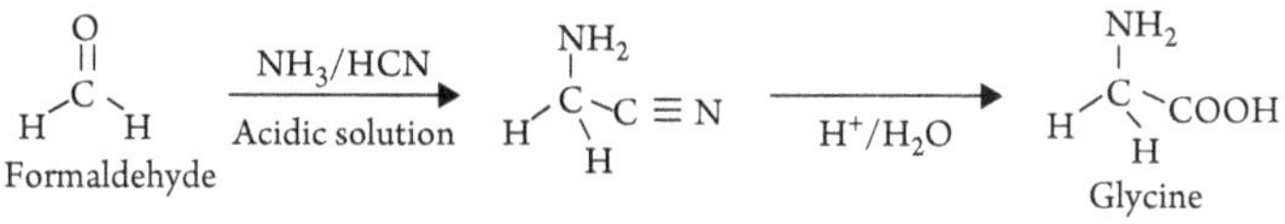

Figure 4.2. The basic steps of the Strecker synthesis.

of gases simulating those of the gas giants, and Miller convinced Urey to let him try an experiment in which a mixture of methane, hydrogen, ammonia, and water vapor was exposed directly to an electrical discharge simulating the effect of lightning. Miller's intuitive expectation was that the energy of the discharge would activate reactions and lead to more complex molecules. His expectation was dramatically fulfilled when he found evidence that two amino acids—glycine and alanine—were among the products (Miller, 1953). Miller's sole-author paper was published in *Science* in 1953 and is now cited as the breakthrough that launched experimental research on the origin of life. The experiment has been repeated many times in later years, with increasingly sensitive analytical instruments, and it is now known that 23 amino acids are synthesized, ten of which are present in proteins.

Miller realized that he was seeing a version of the Strecker synthesis (see Fig. 4.2) discovered by Adolph Strecker 100 years earlier in which ammonia and hydrogen cyanide react with aldehydes to produce alpha-aminonitriles. The nitrile can be hydrolyzed to form a carboxylate group, resulting in an amino acid. Strecker used acetaldehyde to synthesize alanine, and in Miller's reaction conditions formaldehyde was also produced resulting in both glycine and alanine.

Nucleobases

A few years later, another startling result was reported, that adenine, one of the bases of nucleic acids, can be synthesized from hydrogen cyanide in alkaline solutions (Oro and Kimball, 1961). Adenine is a pentamer of HCN as illustrated in Figure 4.3.

More recently, a similar reaction was established by Raffael Saladino working in collaboration with Ernesto DiMauro and their colleagues (Saladino et al., 2012), but using formamide rather than cyanide (Fig. 4.4). This may be a more plausible source of nucleobases because formamide is a relatively stable liquid compared to cyanide:

> We report the one-pot synthesis of all the natural nucleobases, of amino acids and of eight carboxylic acids (forming, from pyruvic acid to citric acid, a continuous series encompassing a large part of the extant Krebs cycle). These data shape a general prebiotic scenario consisting of carbonaceous meteorites acting as catalysts and of a volcanic-like environment providing heat, thermal waters and formamide.

Rotelli et al. (2016) also observed that certain mineral structures composed of metal sulfides had a marked catalytic effect on such reactions:

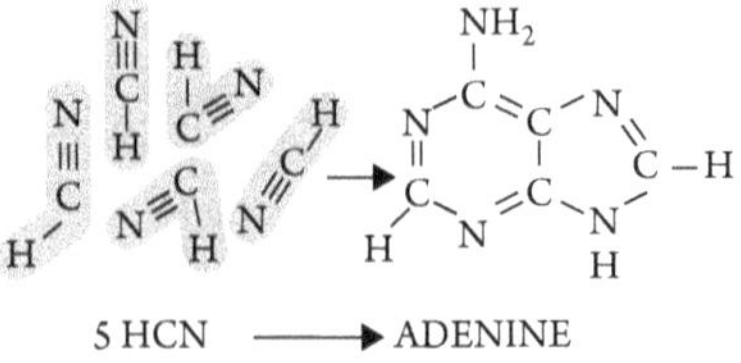

5 HCN ⟶ ADENINE

Figure 4.3. Adenine is a pentamer of HCN, which is a linear molecule. For purposes of illustration, HCN atoms are superimposed on the structure of adenine to show how a seemingly complex molecule can be synthesized from a much simpler reactive molecule. The actual reaction mechanism is more complicated than simply forming bonds between HCN molecules as shown, but adenine does incorporate five carbon and five nitrogen atoms of cyanide.

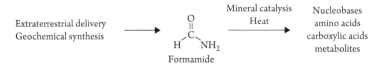

| Extraterrestrial delivery
Geochemical synthesis | ⟶ | Formamide | Mineral catalysis
Heat ⟶ | Nucleobases
amino acids
carboxylic acids
metabolites |

Figure 4.4. Synthesis of biologically relevant compounds from formamide (Saladino et al. 2012).

Here, we show that MSH membranes are catalysts for the condensation of NH_2CHO, yielding prebiotically relevant compounds, including carboxylic acids, amino acids, and nucleobases. Membranes formed by the reaction of alkaline (pH 12) sodium silicate solutions with $MgSO_4$ and $Fe_2(SO_4)_3 \cdot 9H_2O$ show the highest efficiency, while reactions with $CuCl_2 \cdot 2H_2O$, $ZnCl_2$, $FeCl_2 \cdot 4H_2O$, and $MnCl_2 \cdot 4H_2O$ showed lower reactivity. The collections of compounds forming inside and outside the tubular membrane are clearly specific, demonstrating that the mineral self-assembled membranes at the same time create space compartmentalization and selective catalysis of the synthesis of relevant compounds.

Something that is not emphasized in these papers is that actual yields are very low and can only be detected because of the sensitivity of the analytical methods. A typical experiment might begin with 226 mg of formamide, but typical yields are in the range of micrograms. Furthermore, the catalysis by metal sulfides requires a very basic pH range, so even though this reaction can be demonstrated in laboratory conditions it seems unlikely that it would be a major source of organic compounds in the acidic pH range of hydrothermal fields on the early Earth.

Carbohydrates

The formose reaction was discovered by Aleksandr Butlerov (1861), who reported that formaldehyde reacts with itself to produce hundreds of more complex molecules, many of which would be classified as carbohydrates. The reaction is referred to as the formose reaction (a combination of formaldehyde and the suffix -ose used in the terminology of naming sugars) and requires alkaline conditions and a catalyst such as calcium. Because of the robustness of the reaction, there is a consensus that it is a plausible source of carbohydrates in the prebiotic environment.

The formose reaction is initiated when two formaldehyde molecules condense into glycolaldehyde with two carbon atoms (Fig. 4.5); a third formaldehyde is added to produce glyceraldehyde, an intermediate of glycolysis and classified as the simplest three-carbon carbohydrate. The glyceraldehyde can isomerize into dihydroxyacetone which reacts with glycolaldehyde to produce five-carbon sugars called ribulose and ribose. The ribose, of course, is the ribo-constituent of ribonucleic acid, which is why the formose reaction is such a significant aspect of prebiotic chemical evolution. The formose reaction produces hundreds of compounds, with ribose representing just a tiny fraction. However, if borate is present in the alkaline conditions, ribose is stabilized as a borate complex and then accumulates to become a major component of the mixture (Benner et al., 2012)

Given that formaldehyde is essential both for the Strecker synthesis of amino acids and the formose reaction, is it conceivable that there was a continuous source of formaldehyde on the early Earth? A potential answer to this question was reported in 1980 by Yuk Yung and his students at Caltech, who published a paper with the title Photochemical Production of Formaldehyde in Earth's Primitive Atmosphere (Pinto et al., 1980). By analyzing the photochemistry of ultraviolet light exciting carbon dioxide and water vapor in the upper atmosphere, the authors showed that "... formaldehyde could have been produced at a rate of 10^{11} moles per year and then fall to Earth as rain" (Plate 4.6). This fulfills the requirement for a continuous source of a reactive compound which, along with HCN, can lead to all but one of the four primary monomers of life.

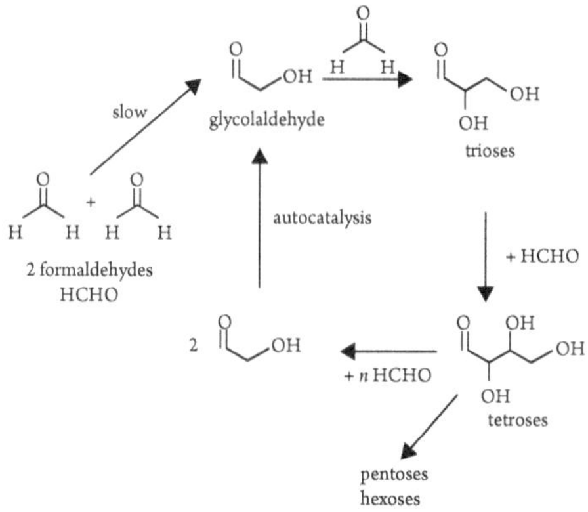

Figure 4.5. Formaldehyde reacts with itself to produce carbohydrates, including ribose and glucose.

Amphiphilic Compounds

Amphiphiles get their name from the fact that they have both a polar or ionic hydrophilic and a nonpolar hydrophobic group on the same molecule. This gives them special properties of self-assembly that will be described in the next chapter. The most common amphiphiles that everyone knows from direct experience are soaps and detergents. A soap is simply a monocarboxylic acid having ten or more carbon atoms in a hydrocarbon chain that ends in a carboxyl group ($-COOH$). The carboxyl group is a weak acid that loses a proton at higher pH ranges to become a carboxylate anion ($-COO^-$). Monocarboxylic acids have two important properties related to their function in living cells today. As triglycerides, they represent stored chemical energy that can be released by oxidation, and as phospholipids, they are the primary components of cell membranes. Given their role in membrane structure, it seems likely that monocarboxylic acids also were essential components of the membranes of the first forms of cellular life. But what sources were possible?

One such source is a reaction first described in 1925 by Franz Fischer and Hans Tropsch, who discovered that a mixture of carbon monoxide and hydrogen gas passed over a hot iron catalyst produced long-chain aliphatic alkanes and their derivatives (Anderson, 1984). Figure 4.6 shows a simplified version of the main steps involved in the Fischer-Tropsch (FT) synthesis of hydrocarbon chains. The gray spheres symbolize the atoms of a metal surface such as iron. When heated and exposed to a mixture of carbon monoxide (CO) and hydrogen gas (H_2), the CO adsorbed to the surface undergoes successive cycles of reduction, dehydration, and carbon–carbon bond formation to produce a growing hydrocarbon chain.

John Oro decided that it would be worth testing whether the FT reaction could synthesize hydrocarbon chains in a laboratory simulation of prebiotic conditions. The procedure was remarkably simple, involving a mixture of hydrogen and carbon monoxide passed over a hot iron catalyst. However, this was not just ordinary iron. Instead, Oro and his students used an iron-nickel powder prepared by grinding a sample of the Canyon Diablo meteorite that impacted the Arizona desert 40,000 years ago and left a crater nearly a kilometer in diameter. The tube containing the powder was heated, the gas mixture was passed through, and the vapor coming out the other end of the tube was

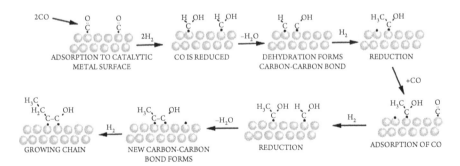

Figure 4.6. The main steps involved in the Fischer-Tropsch synthesis of hydrocarbon chains.

analyzed by gas chromatography. Just as Fischer and Tropsch would have predicted, the vapor contained a mixture of alkanes, long-chain monocarboxylic acids and alcohols (Nooner and Oro, 1980).

Later research by Bernd Simoneit and his students used an even simpler simulation in which oxalic acid was exposed to moderately elevated temperatures and pressures in a closed steel container often referred to as a "bomb." Under these conditions, the compounds dissociated into a mixture of hydrogen gas, carbon dioxide, and carbon monoxide. After several hours of heating at temperatures of 200° C, the bomb was cooled and the contents analyzed (McCollom et al., 1999; Rushdi and Simoneit, 2001). The products were again a mixture of alkanes, long-chain monocarboxylic acids and alcohols similar to that reported by Oro.

Simoneit et al. (2007) also found that if stoichiometric glycerol was present in the mixture along with a variety of fatty acids, the same conditions produced good yields of monoglycerides. These results are significant because the majority of membrane-forming lipids today are glycerol esters. Furthermore, monoglycerides by themselves can assemble into lipid bilayers and when mixed with fatty acids, are able to form stable membranes (Monnard et al., 2002; Mansy and Szostak, 2008; Maurer et al., 2009).

There are two reasons why these results are significant for our understanding of the origin of life. The first is that alkanes and monocarboxylic acids are present in the organic mixtures in carbonaceous meteorites, and it is not difficult to imagine that Fischer-Tropsch-type (FTT) synthesis could have occurred in the parent bodies as they heated up in the early solar system, as suggested by Ed Anders in 1963. Second, and even more important, FTT synthesizes amphiphilic compounds that can self-assemble into membranous compartments, as will be described in the next chapter. This means that such compounds were likely to be components of the prebiotic organic mixture available on the early Earth to play a role in the steps leading toward the origin of life.

Mineral Interface Simulations

Mineral surfaces have long been a focus of attention in origins of life research, and a recent review by Hazen (2017) provides a thoughtful perspective on this important aspect of prebiotic chemistry. The interest in minerals began with Bernal's speculation (1951) that the absorptive properties of clay may have concentrated potential reactants and thereby promoted reactions not otherwise possible in dilute solutions. Ferris (2006) devised experimental models of clay systems and established that the concentrating effect could in fact promote polymerization of activated nucleotides, a reaction that will be described in detail in Chapter 9. However, here we are focusing on synthesis of biologically relevant monomers such as amino acids and potential metabolites driven by interactions with mineral surfaces.

One such reaction was proposed by Günther Wächtershäuser (1990) who noted that when iron and sulfide reacted to produce the mineral pyrite (FeS_2) a potential source of reducing power was made available. From this observation Wächtershäuser developed an elaborate vision which involved reduction of carbon dioxide to intermediates that were incorporated in a primitive metabolism associated with the mineral surface. To test the idea, Wächtershäuser collaborated with Claudia

Huber, an organic chemist at the Technical University in Munich, who simulated the conditions of a black smoker hydrothermal vent in which iron and sulfide are prominent solutes in the hot water emerging from the vents and become components of the mineral chimneys (Huber and Wachtershauser, 1997). Huber devised a reactor in which iron and nickel sulfides were heated to boiling temperatures in the presence of carbon monoxide. They observed that carbon–carbon single bonds were synthesized to produce acetic acid and its thioester, which led them to conclude that: "The results support the theory of a chemoautotrophic origin of life with a CO-driven, (Fe,Ni)S-dependent primordial metabolism."

Similar ideas were developed by Michael Russell, William Martin, and Nick Lane, who suggested in a series of papers that the dissolved hydrogen gas available in hydrothermal vents could reduce carbon dioxide to a series of biologically relevant compounds that then evolved into a primitive version of metabolism. Their ideas will be described in Chapters 7 and 8.

Primitive Metabolism: Can Geochemical Reactions Prime the Metabolic Pump?

George Cody and Robert Hazen suggested that not only could organic compounds be synthesized geochemically on the early Earth, but they could also be incorporated into a system of reactions proposed by Harold Morowitz. In his book *Beginnings of Cellular Life* (2004), Morowitz summarized the basic idea that we can deduce the chemical components of a primitive metabolism from those involved in metabolism today, particularly glycolysis and the citric acid cycle. A keystone metabolite of these pathways is pyruvic acid, so the prediction is that it may be possible to discover a source of this compound that would be plausibly available in the prebiotic environment. Cody and Hazen (with others) established a geothermal simulation in which carbon monoxide interacts with a hot iron sulfide mineral at a pressure several hundred times higher than that of our atmosphere (Cody et al., 2001). Similar to the FT reaction, the carbon monoxide was adsorbed to the mineral surface and reacted to produce multiple organic compounds, one of which was pyruvic acid. The authors concluded that:

> The natural synthesis of such compounds is anticipated in present-day and ancient environments wherever reduced hydrothermal fluids pass through iron sulfide-containing crust .... These compounds could have provided the prebiotic Earth with critical biochemical functionality.

Plate 4.7 illustrates how pyruvic acid, a relatively simple organic molecule, can be transformed by pressure and elevated temperature into more complex products with physical properties of self-organization (Hazen and Deamer, 2007). When pyruvic acid was heated to 200° C in a gold capsule for two hours, thousands of new compounds were produced, some with amphiphilic properties and the ability to organize into compartments as shown in Plate 4.7 by phase microscopy (left), crossed polarized filters (center), and fluorescence (right). It's remarkable that the mass of polymeric material pictured began as crystalline, water-soluble, pyruvic acid, a compound that is at the center of metabolism in all life today.

Prebiotic Synthesis of Nucleotides

The nucleotide monomers of nucleic acids consist of a purine or pyrimidine linked through a nitrogen–carbon bond to a pentose sugar, which in turn is phosphorylated through an ester bond on the 5' hydroxyl group. Each of the component molecules of nucleotides was presumably present in organic mixtures on the prebiotic Earth, but they must be linked into the more complex molecular structure of nucleotides before they can be incorporated into nucleic acid polymers. One might imagine that phosphorylation of a ribose sugar could occur because ester bonds are not difficult to synthesize by simple condensation reactions, but synthesis of the C–N bond between a sugar and a nucleobase has been much more challenging.

A paper by Powner et al. (2009) reported a series of reactions that led to mononucleotide synthesis. Instead of attempting to produce C–N bonds between an existing pyrimidine such as cytosine and ribose, the reaction used sequential additions of cyanamide, cyanoacetylene, glycolaldehyde, and glyceraldehyde, all in the presence of phosphate. A later paper (Powner et al., 2010) extended this approach to a more general synthesis of organic compounds relevant to the origin of life. Although it is unlikely that such complex mixtures of reactants might have been present in the prebiotic environment, the fact that the reaction can occur in aqueous solution, using only the chemical energy of the reactants, opens a new direction for future investigations that may reveal simpler processes.

Decomposition and Steady States

There is an understandable tendency to emphasize synthetic reactions in origins of life research, and this is why the results we have discussed so far have been impressive. What is often lacking in descriptions of prebiotic chemical evolution is the inconvenient fact that most of the synthesized compounds have limited lifetimes in aqueous solutions. Some undergo hydrolysis reactions in which the addition of water to a chemical bond breaks the bond, the rate of which can be vastly affected by temperature. For instance, Stockbridge et al. (2010) pointed out that the hydrolysis of phosphodiester bonds like those in nucleic acids increases more than 27,000-fold when temperature increases from 25 to 100° C. Does this mean that nucleic acids could not exist in hydrothermal fields? Although they would decompose at much faster rates at elevated temperatures, the phosphodiester bonds still have an estimated half-life of many years.

Deamination occurs when an amine group ($-NH_2$) is replaced by a hydroxyl group ($-OH$) on the cytosine ring with the product being uracil. Shapiro (1999) noted that this leads to a problem with theories of the origin of the genetic code. If cytosine quickly decomposes to uracil after it is synthesized, how can a genetic code evolve when only three bases are available? A related hydrolysis reaction is depurination of the adenine and guanine nucleotides in nucleic acids, in which the purine detaches from the ribose or deoxyribose ring by hydrolysis of the N–glycoside bond.

Another decomposition reaction is decarboxylation of amino acids to form amines that can no longer undergo polymerization. The aptly named putrescine and cadaverine result when arginine and lysine undergo decarboxylation to form products having four and five carbon chains with an amine group at either end. These diamines with a characteristic odor are produced when bacteria use decaying proteins as a source of nutrients.

The most damaging decomposition process is called pyrolysis, a term derived from Greek words meaning "fire" and "breakage." When any bioorganic compound is heated much above 200° C, the chemical bonds holding it together begin to break and release simpler compounds as products. Miller and Bada (1988) pointed out that amino acids cannot survive the temperatures associated with black smokers and that this would put an upper limit on the origin of life in those conditions.

Other sources of degradation are random crosslinking reactions that produce a solid polymer called kerogen when sedimentary biological material is heated to temperatures ranging from 200 to 500° C. Kerogen is found in coal and shales and is also a major component of the organic material in carbonaceous meteorites. During the heating process, lipids in the material are relatively stable and ultimately end up as liquid hydrocarbons called petroleum. The material left behind then undergoes increasing cross linking that includes loss of hydrogen, oxygen, and nitrogen. As this occurs, the material becomes increasingly aromatic so that kerogens are generally defined as a complex polymerized mixture of polycyclic aromatic hydrocarbon rings. The end point of the process is graphite composed of sheets of carbon atoms bonded in hexagonal arrays.

Summary and Open Questions

This chapter described four levels of increasing complexity, some of which are well understood while others have open questions yet to be addressed. The synthesis of CONPS, the biogenic elements, by nuclear fusion in stars is well understood, as is the composition of molecular products that accumulate on IDPs and are then incorporated into comets and asteroids. We also know from laboratory simulations that those same reactive molecules can be synthesized by reactions that would likely occur through geochemical and photochemical reactions on the Earth's surface. An important open question is the relative amounts supplied by delivery as infall on IDPs, comets, and meteorites, and how much is synthesized by geochemical reactions.

A related question concerns how the organic material is processed by the conditions of the Hadean era. Some was certainly degraded by pyrolysis and hydrolysis, but at the same time others were being synthesized. This is referred to as turnover in steady state conditions and so far has attracted only minimal interest.

Finally, it is clear that in order to react, the compounds must be sufficiently concentrated. There are three possibilities. The most obvious is evaporation, but adsorption to mineral surfaces has also been proposed. Thermal gradients existing in porous media have also been demonstrated to concentrate solutes (Budin et al., 2012; Herschy et al., 2014)

The point of this discussion is that the organic compounds in the prebiotic environment were not simply a stable pool of the primary small molecule components of life. Instead there must have been a continuing turnover process in which organic material was either delivered or synthesized then degraded and recycled in a variety of ways. A small fraction of this organic mixture was soluble in water and happened to have properties that allowed it to undergo chemical and physical processes relevant to the origin of life.

A volcanic setting is a plausible site for the chemical and physical events leading up to the origin of cellular life. But instead of the interface between water and mineral surfaces it seems more likely to me that the fluctuating conditions characteristic of hydrothermal springs will be a more fruitful site to model experimentally. Only under those particular conditions do we have a combination of concentrating effects, mineral surfaces that can act as organizing agents, and free energy available to drive the reactions required to assemble the first protocells.

Now we come to one of the many gaping holes in the story that will be discussed further in Chapter 10. We know that the main components of life can be chemically synthesized in laboratory simulations of prebiotic chemistry and that this can occur outside the laboratory is confirmed by their representation among the organic compounds present in carbonaceous meteorites. This is a satisfying step toward understanding how life can begin, but three major problems emerge. First, there are thousands of compounds present in the mixture of organics, whether they are delivered or synthesized. How could they possibly be sorted out into just those required for the origin of life? Second, because they would be present as very dilute solutions, how could these compounds be concentrated sufficiently to begin to react with one another? Third, even though we know of synthetic reactions that produce biologically relevant products, we also know that a series of decomposition reactions will continuously degrade such molecules into inert products that cannot participate in useful reactions. How can prebiotic chemistry attain a steady state in which synthesis and decomposition rates are balanced?

In the chapters to follow, we will propose speculative answers to these questions but will offer a brief sketch here. The easiest to address is the concentration question, and by far the most plausible way to do so is also the most obvious. If the organic solutes are in a freshwater system undergoing cycles of evaporation and rehydration, the solutes continuously cycle between extremely concentrated films on mineral surfaces and back to dilute solutions upon rehydration. The second question can also be answered by reference to self-assembly and relative reactivities. For instance, no matter how complex a mixture of organics might be, amphiphilic compounds are readily separated from the mixture by the process of self-assembly into relatively pure components of monolayers, micelles, and membranes. Self-assembly of purified material also brings together concentrated reactants so that specific chemical reactions can occur. Only those compounds that can react will react, while chemically inert compounds will be left behind.

Setting the Stage

1. Nuclear fusion produces elemental carbon, silicon, oxygen, nitrogen, phosphorus, and sulfur in the hot, late stages of an ordinary star's life.

2. The star sheds much of its mass in the process, producing the silicate mineral dust that forms immense molecular clouds throughout our own galaxy and neighboring galaxies like Andromeda.
3. The surfaces of dust particles collect thin icy films of the most common molecules in the interstellar medium, including H_2O, CO, CO_2, CH_3OH, and NH_3.
4. As new stars appear in the molecular clouds, their UV radiation drives photochemical reactions in the mixed ice, and a variety of more complex organic compounds are produced.
5. The dust particles aggregate into planetesimals that ultimately accrete into planets, and their coat of mixed ice is incorporated along with its content of organic compounds.
6. Asteroids and comets are planetesimals that escaped planet formation, and carbonaceous meteorites are samples of the surfaces of asteroids. The organic compounds in such meteorites include amino acids, purines and pyrimidines, long-chain monocarboxylic acids, and simple sugars.
7. Dust particles, now called IDPs, are the main mineral mass delivered continuously to the Earth after the moon-forming event.
8. When IDPs enter the atmosphere at high velocity, their mixed ice coating and embedded organics are shed within milliseconds as they heat up and then cool. The naked IDPs drift to the Earth's surface, and their content of water and organics falls as precipitation.
9. The rain and solutes mostly enter the ocean where organic compounds are diluted to micromolar concentration. Organics that fall on land accumulate and then are flushed into hydrothermal pools. When the pools undergo cycles of evaporation, the organic solutes become highly concentrated as films on mineral surfaces.
10. Organic compounds are also being synthesized in the atmosphere by photochemical reactions and by geochemical processes associated with volcanism. These also accumulate in films on mineral surfaces, presumably in mixtures with organic compounds delivered by IDPs. The reactions that occur in concentrated films of organic compounds are the topic of the chapters to follow.

References

Allamandola LJ, Sandford SA, Valero GV (1988) Photochemical and thermal evolution of interstellar/precometary ice analogs. *Icarus* 74, 225–252.

Anders E (1989) Pre-biotic organic matter from comets and asteroids. *Nature* 342, 255–257.

Anderson RB (1984) *The Fischer-Tropsch Reaction*. London: Academic Press.

Bada JL, Lazcano A (2003) Prebiotic soup—revisiting the Miller experiment. *Science* 300, 1745–746.

Benner SA, Kim H-J, Carrigan MA (2012) Asphalt, water, and the prebiotic synthesis of ribose, ribonucleosides, and RNA. *Acc Chem Res* 45, 2025–2034.

Bernal JD (1951) *The Physical Basis of Life*. London: Routledge and Kegan Paul,.

Budin I, Debnath A, Szostak JW (2012) Concentration-driven growth of model protocell membranes. *J Am Chem Soc* 134, 20812–20819.

Butlerow A (1861) Formation of a sugar-like substance by synthesis (translated from German). Justus Liebigs *Annalen der Chemie*, 120, 295–298.

Chyba C, Sagan C (1992) Endogenous production, exogenous delivery and impact-shock synthesis of organic molecules: An inventory for the origins of life. *Nature* 355, 125–132.

Clemett SJ, Maechling CR, Zare RN, Swan PD, Walker RM (1993) Identification of complex aromatic molecules in individual interplanetary dust particles. *Science* 262,721–725.

Cody GD, Boctor NZ, Hazen RM, Brandes JA, Morowitz HJ, Yoder HS (2001) Geochemical roots of autotrophic carbon fixation: Hydrothermal experiments in the system citric acid, H_2O-(±FeS)–(±NiS). *Geochim Cosmochim Acta* 65, 3557–3576.

Delsemme AH (2000) Cometary origin of the biosphere. *Icarus* 146, 313–325.

Dworkin JP, Deamer DW, Sandford SA, Allamandola LJ (2001) Self-assembling amphiphilic molecules: Synthesis in simulated interstellar/precometary ices. *Proc Natl Acad Sci USA* 98, 815–819.

Ehrenfreund P, Cami J (2010) Cosmic carbon chemistry: From the interstellar medium to the early Earth. *CSH Perspectives in Biology.* doi: 10.1101/cshperspect.a002097.

Ferris JP (2006) Montmorillonite-catalysed formation of RNA oligomers: The possible role of catalysis in the origins of life. *Phil Trans Royal Soc B* doi: 10.1098/rstb.2006.1903.

Garrod RT, Widicus Weaver SL, Herbst E (2008) Complex chemistry in star-forming regions: An expanded gas-grain warm-up chemical model. *Astrophys J* 682, 283–302.

Glavin DP, Bada JL. (2004) Survival of amino acids in micrometeorites during atmospheric entry. *Astrobiology* 1: 259–269.

Greenberg JM (2002) Cosmic dust and our origins. *Surface Science* 500, 793–822.

Hazen RM (2017) Chance, necessity, and the origins of life. *Phil Trans Royal Society A* doi: 10.1098/rsta.2016.035.

Hazen RM, Deamer DW (2007) Hydrothermal reactions of pyruvic acid: Synthesis, selection, and self-Assembly of amphiphilic molecules. *Orig Life Evol Biosph* 37, 143–152.

Herschy B, Whicher A, Camprubi E, Watson C, Dartnell L, Ward J, Evans JR, Lane N (2014) An origin-of-life reactor to simulate alkaline hydrothermal vents. *J Mol Evol* 79, 213–227.

Huber C, Wächtershäuser G (1997) Activated acetic acid by carbon fixation on (Fe,Ni)S under primordial conditions. *Science* 276, 245–247.

Love SG, Brownlee DE (1991) An interplanetary dust particle linked directly to type CM meteorites and an asteroid origin. *Science* 251:549–552.

Mansy SS, Szostak JW (2008) Thermostability of model protocell membranes. *Proc Natl Acad Sci U S A.* 105, 13351–13355.

Martins Z, Botta O, Fogel ML, Sephton MA, Glavin DP, Watson JW, Dworkin JP, Schwartz AW, Ehrenfreund P (2008) Extraterrestrial nucleobases in the Murchison meteorite. *Earth Planetary Sci Lett* 270, 130–136.

Matrajt, G. Walker R, Brownlee D, Joswiak D (2011) Diverse forms of primordial organic matter identified in interplanetary dust particles. *Meteorit Planet Sci* doi: 10.1111/j.1945-5100.2011.01310.x.

Maurer SE, Deamer DW, Boncella JM, Monnard PA (2009) Chemical evolution of amphiphiles: Glycerol monoacyl derivatives stabilize plausible prebiotic membranes. *Astrobiology* 9, 979–987.

McCollom TM, Ritter G, Simoneit BRT (1999) Lipid synthesis under hydrothermal conditions by Fischer-Tropsch-type reactions. *Orig Life Evol Biosph* 29, 153–166.

Messenger S, Nakamura-Messenger K, Keller LP, Clemett SJ (2015) Pristine stratospheric collection of interplanetary dust on an oil-free polyurethane foam substrate. *Meteorit Planet Sci* 50, 1468–1485.

Miller S (1953) A production of amino acids under possible primitive Earth conditions. *Science* 117, 528–529.

Miller SL, Bada JL (1988) Submarine hot springs and the origin of life. *Nature* 334, 609–611.

Monnard P-A, Apel CL, Kanavarioti A, Deamer DW (2002) Influence of ionic inorganic solutes on self-assembly and polymerization processes related to early forms of life: Implications for a prebiotic aqueous medium. *Astrobiol* 2, 139–152.

Morowitz HJ (2004) *Beginnings of Cellular Life: Metabolism Recapitulates Biogenesis.* New Haven, CT: Yale University Press.

Mumma MJ, Charnley SB (2011) The chemical composition of comets: Emerging taxonomies and natal heritage. *Ann Rev Astron Astrophys* 49, 471–524.

Nooner DW, Oro J (1980) Synthesis of fatty acids by a closed system Fischer-Tropsch *Process Adv Chem* 178, 159–171.

Oro J (1961) Comets and the formation of biochemical compounds on the primitive Earth. *Nature* 190, 389–390.

Oro J, Kimball AP (1961) Synthesis of purines under possible primitive earth conditions. I. Adenine from hydrogen cyanide. *Arch Biochem Biophys* 94, 217–227.

Pearce BKD, Pudritz RE, Semenov DA, Henning TK (2017) Origin of the RNA world: The fate of nucleobases in warm little ponds. *Proc Natl Acad Sci USA* 114, 11327–11332.

Pinto JP, Gladstone R, Yung YL (1980) Photochemical production of formaldehyde in Earth's primitive atmosphere. *Science* 210, 183–185.

Pizzarello S, Shock E. (2010) The organic composition of carbonaceous meteorites: The evolutionary story ahead of biochemistry. *Cold Spring Harb Perspect Biol* doi: 10.1101/cshperspect.a002105.

Pizzarello S, Williams LB, Lehman J, Holland GP, Yarger JL (2011) Abundant ammonia in primitive asteroids and the case for a possible exobiology. *Proc Natl Acad Sci USA* 108, 4303–4306.

Powner MW, Gerland B, Sutherland JD (2009) Synthesis of activated pyrimidine ribonucleotides in prebiotically plausible conditions. *Nature* 459, 239–242.

Powner MW, Sutherland JD, Szostak JW (2010) Chemoselective multicomponent one-pot assembly of purine precursors in water. *J Am Chem Soc* 132, 16677–16688.

Rotelli L, Trigo-Rodriguez JM, Moyano-Cambero CE, Saladino R (2016) The key role of meteorites in the formation of relevant prebiotic molecules in a formamide/water environment. *Scientific Reports* 6, 38888.

Rushdi AI, Simoneit BRT (2001) Lipid formation by aqueous Fischer-Tropsch-type synthesis over a temperature range of 100 to 400° C. *Orig Life Evol Biosph* 31, 103.

Saladino R, Crestini C, Pino S, Costanzo G, Di Mauro E (2012) Formamide and the origin of life. *Phys Life Rev* 9, 84–104.

Shapiro R (1999) Prebiotic cytosine synthesis: A critical analysis and implications for the origin of life. *Proc Natl Acad Sci USA* 96, 4396–4401.

Simoneit BRT, Rushdi AI, Deamer DW (2007) Abiotic formation of acylglycerols under simulate hydrothermal conditions and self-assembly of such lipid products. *Adv Space Res* 40, 1649–1656.

Stockbridge RB, Lewis CA, Yuan Y, Wolfenden R (2010) Impact of temperature on the time required for the establishment of primordial biochemistry, and for the evolution of enzymes. *Proc Natl Acad Sci USA* 107, 22102–22105.

Wächtershäuser G (1990) Evolution of the first metabolic cycles. *Proc Natl Acad Sci USA* 87, 200–204.

Wing MR, Bada JL (1991) Geochromatography on the parent body of the carbonaceous chondrite Ivuna. *Geochim Cosmochim Acta* 55, 2937–2942.

5

Self-Assembly Processes Were Essential for Life's Origin

> Life began with little bags of garbage, random assortments of molecules doing some crude kind of metabolism . . . the garbage bags grow and occasionally split in two, and the ones that grow and split fastest win.
> Freeman Dyson, 1999

Overview and Questions to be Addressed

In the absence of self-assembly processes, life as we know it would be impossible. This chapter begins by introducing self-assembly then focuses on the primary functions of membranes in living cells, most of which depend on highly evolved proteins embedded in lipid bilayers. These serve to capture light energy in photosynthesis and produce ion concentration gradients from which osmotic energy can be transduced into chemical energy. Although lipid bilayer membranes provide a permeability barrier, they cannot be absolutely impermeable because intracellular metabolic functions depend on external sources of nutrients. Therefore, another set of embedded proteins evolved to form transmembrane channels that allow selective permeation of certain solutes.

The earliest life did not have proteins available, so in their absence what was the primary function of membranous compartments in prebiotic conditions? There are three possibilities. First, the compartments would allow encapsulated polymers to remain together as random mixtures called protocells. Second, populations of protocells that vary in composition would be subject to selective processes and the first steps of evolution. Even though any given protocell would be only transiently stable, certain mixtures of polymers would tend to stabilize the surrounding membrane. Such an encapsulated mixture would persist longer than the majority that would be dispersed and recycled, and these more robust protocells would tend to emerge as a kind of species.

Last and perhaps most important, there had to be a point in early evolution at which light energy began to be captured by membranous structures, just as it is today. Bilayer membranes are not necessarily composed solely of amphiphilic molecules. They can also contain other nonpolar compounds that happen to be pigments capable of capturing light energy. This possibility is almost entirely unexplored, but the experiments are obvious and would be a fruitful focus for future research.

Questions to be addressed:

- What is meant by self-assembly?
- Why is self-assembly important for the origin of life?
- What compounds can undergo self-assembly processes?
- How can mixtures of monomers and lipids assemble into protocells?

Self-Assembly by Base Pairing and Base Stacking

We tend to think of living cells in terms of directed assembly. This process incorporates what Francis Crick called the central dogma: genetic information stored in deoxyribonucleic acid (DNA) is first transcribed into a coded sequence of bases in messenger RNA (mRNA), then delivered to ribosomes where sequences of base triplets in the mRNA called codons direct the incorporation of amino acids into proteins. However, embedded in the life process are more subtle effects referred to as self-assembly which are physical rather than chemical in nature (Fig. 5.1). Before the first forms of life evolved the ability to direct the assembly of functional polymers like proteins and nucleic acids, the processes of self-assembly produced structures that were steps in the pathway to life and in fact were essential for bringing together the molecular systems that ultimately found ways to grow and reproduce.

The most familiar self-assembly forces are the hydrogen bonds that stabilize base pairing in nucleic acids. If the base pairs of the DNA double helix are disrupted, for instance simply by heating, the DNA "melts" and separates into single strands. When cooled, the single strands spontaneously reassemble by base pairing to form the original double helix. The same base-pairing forces also stabilize the specific structures of transfer RNA (tRNA) and the hairpins that form in mRNA and ribosomal RNA (rRNA).

A more general force that stabilizes the structure of certain single-stranded nucleic acids and the double helix of DNA is the stacking of nucleotide bases within the strand. The attraction results from the interaction of pi electrons in the aromatic rings of neighboring nucleobases. These have been studied extensively in the DNA double helix, and the surprising result is that base stacking is more significant than base pairing in stabilizing the duplex structure (Yakovchuk et al., 2006). The significance in relation to this book is that nucleobase compounds such as nucleoside monophosphates are assumed to have been present in the prebiotic environment; when concentrated these would tend to form linear structures stabilized by base stacking. As will be described in Chapter 9, such linear structures can be considered to be prepolymers capable of being linked by ester-bond synthesis into polymers required for the origin of RNA-based life.

There is one more self-assembly process to be noted, which is that double-stranded (dsDNA) itself can assemble into a liquid crystalline phase (Fraccia et al., 2015). As will be discussed in Chapter 9, the organized matrix of pure DNA can also promote a ligation reaction having relevance to life's origins.

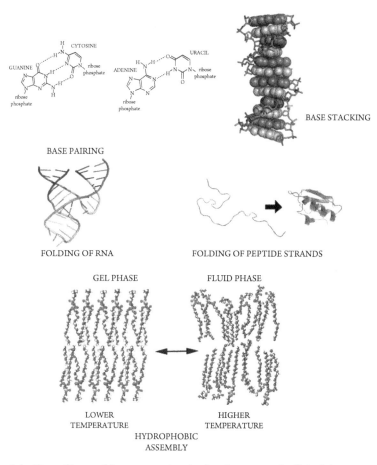

CYTOSINE

GUANINE   ribose phosphate

URACIL

ADENINE   ribose phosphate

ribose phosphate

ribose phosphate

BASE STACKING

BASE PAIRING

FOLDING OF RNA   FOLDING OF PEPTIDE STRANDS

GEL PHASE   FLUID PHASE

LOWER   HIGHER
TEMPERATURE   TEMPERATURE
HYDROPHOBIC
ASSEMBLY

Figure 5.1. Five self-assembly processes involved in the origin of cellular life.

Self-Assembly by Folding

The next self-assembly process relevant to the origin of life is that longer strands of biopolymers like nucleic acids and proteins do not exist in solution as threads. Instead there is a strong tendency to form globular structures stabilized by hydrogen bonding, electrostatic interactions, and entropic effects. An important point is that random sequences of monomers—nucleotides and amino acids in the strands—may undergo folding but the resulting structures are all different and generally lack functional capacity. However, even in random sequences a few rare structures will happen to exhibit functions such as catalytic activity and would be classified as ribozymes (RNA) or enzymes (proteins). *The central theme of this book is that these rare polymers can only become functional systems if they are encapsulated in a membranous compartment. The system and the compartment can then be subjected to selective processes that drive the first steps of evolution.* A proposal for how this can happen is described in Chapters 9 and 10.

Self-Assembly by Hydrophobic Interactions of Hydrocarbon Chains

The third process essential for life is the self-assembly of certain lipids into membranous boundary structures (Walde et al. 1994). The stability of such membranes is largely due to the hydrophobic effect, first defined by Charles Tanford (1973). The effect emerges from two molecular forces, the first one related to the entropy of the system of lipids in an aqueous phase and the second to van der Waals interactions between the hydrocarbon chains. These forces will be described in further detail later in this chapter.

The main point to be made here is that some of the most important first steps toward the origin of life in prebiotic conditions would have been due to self-assembly of organic compounds dispersed or dissolved in water. These are understood in terms of spontaneous physical processes rather than being primarily chemical in nature. In a sense, only after self-assembly of certain structures had occurred could the chemical reactions essential for life begin.

Structure of Cell Membranes

Helpful clues to understanding how amphiphilic compounds were involved in the origin of life are provided by the physical and chemical properties of lipids that compose the membranes of all life today. Lipids are generally defined as various biochemicals that are soluble in organic solvents. Examples include fats (triglycerides), fatty acids, phospholipids, and sterols such as cholesterol. In terms of chemical structure, lipids are amphiphilic compounds that have both a hydrocarbon moiety and a hydrophilic group on the same molecule. Virtually all lipids have a hydrophilic group such as carboxylate, phosphate, or hydroxyl attached to one end of the molecule. These groups are polar or ionic and therefore interact strongly with water. The combination of a hydrophobic hydrocarbon and a hydrophilic group on the same molecule is the reason why lipids are called amphiphiles, a term derived from Greek words meaning "loving both." Their amphiphilic character gives lipids unique physical properties that are essential to all life, and a fundamental concept that is the focus of this book is that amphiphilic compounds were also essential for life to begin.

A good place to start the discussion of lipid self-assembly is to briefly recount the history of research on membranes, which also provides a foundation for topics to come later. First, there are four terms that are important to understand: osmosis, hypertonic isotonic, and hypotonic. Osmosis refers to what happens when a membrane is permeable to water molecules but not permeable to a larger solute, such as sucrose, that is more concentrated on one side of the membrane than the other. That concentrated solution is called hypertonic; the more dilute solution is called hypotonic. In that situation, water will tend to move through the membrane until the concentration of solutes is the same on both sides of the membrane, or isotonic.

Some of the earliest studies related to cell membranes were undertaken by Earnest Overton who investigated osmotic effects on cells (Overton, 1895). He observed that cells in plant tissues shrink in volume in hypertonic solutions and concluded that this was best explained if the cells had a semipermeable membrane boundary that was permeable to small water molecules but not to ions or larger molecules like sucrose. At about the same time, Alfonso Luis Herrera, in Mexico, defined plasmogeny as the study of the origin of the protoplasm and claimed that life could be understood according to the laws of chemistry and physics, such as osmosis (see Mendoza, 1995). Furthermore, he demonstrated that certain physical properties of the protoplasm can be simulated with organic and inorganic compounds. For instance, he experimented with a mixture of olive oil and sodium hydroxide and was fascinated by the motions of the droplets that formed and budded off membranous cell-like structures.

The next step toward understanding membrane structure was reported by Gorter and Grendel (1925) who extracted lipid from red cells and spread it as a monolayer on water. The area of the monolayer was twice the calculated surface area of the red cells, and they concluded that the lipid was present as a bimolecular layer surrounding each red cell. In the 1930s, Danielli and Davson (1935) integrated multiple sources of information and concluded that the plasma membrane of cells was in fact a lipid bilayer with proteins attached to its surface.

In the 1950s, the invention of the electron microscope was revolutionary: for the first time membranes could be resolved—after fixing and staining them with osmium tetroxide, embedding them in a plastic, and slicing them into extremely thin sections just 50 - 100 nm thick. Robertson (1967) used electron microscopy to study a variety of cells and reported that not just the cytoplasm was enclosed in a membrane, but intracellular organelles like the nucleus, mitochondria, and endoplasmic reticulum were also essentially membranous structures. Robertson proposed a membrane structure consisting of lipid bilayers with proteins attached to the surface.

The fluid-mosaic membrane proposed by Singer and Nicolson (1972) integrated electron microscopy images with certain membrane functions and is now the consensus model. Fluid lipid bilayers provide a boundary that separates the internal compartment of a cell from the external medium, but most of the functional membrane-associated proteins are embedded in the bilayer rather than simply adhering to the surface as suggested by Robertson.

Properties of Amphiphilic Compounds

By the early 1960s, it had been known for years that a phospholipid called lecithin could be extracted from a variety of sources such as soy beans and egg yolks. It was also known that the addition of water to a few micrograms of dry lecithin on a microscope slide produced worm-like structures, called myelin figures because of their resemblance to the myelin surrounding axons in the nervous system (Plate 5.1).

In the 1960s, Alec Bangham and his coworkers at the Animal Physiology Institute in Babraham, near Cambridge, England, discovered that a milky dispersion of lecithin was produced when dilute salt solutions were added to dry lecithin (Bangham et al., 1965). Under the microscope, it could be seen that the turbidity was caused by microscopic spherical globules. Furthermore, the globules responded to osmotic pressure by shrinking in volume in concentrated, hypertonic solutions or swelling in dilute, hypotonic solutions, just as Overton had demonstrated for the membranes surrounding plant cells. Bangham used an electron microscope to study the organization of lipids within the globules and found that they were composed of multilamellar lipid bilayers. These observations clearly demonstrated that self-assembled lipid bilayers are the primary boundary of cell membranes. In the next section, we will describe the chemical structures of the amphiphilic compounds and the physical properties that allow them to assemble into membranes.

Varieties of Amphiphilic Molecules

The simplest amphiphiles classified as lipids are fatty acids composed of a hydrocarbon chain with a carboxyl group at the end. Some fatty acids have straight chains while others have one or more double bonds (Plate 5.2). Although fatty acids can assemble into membranes under certain conditions (Hargreaves and Deamer, 1978; Chen and Szostak, 2004; Mansy and Szostak 2008) the membranes are relatively fragile and are only stable under certain defined conditions of pH and ionic composition of the medium. This fact will become important later when we consider whether life began in salty seawater or freshwater ponds associated with volcanic land masses on the early Earth.

At some point early in evolution, microbial life evolved biochemical pathways that attach two fatty acids to a glycerol phosphate. Having two chains on the same molecule markedly increases membrane stability, and the resulting phospholipids are the major amphiphilic compounds present in all cell membranes today. Plate 5.2 shows several species of phospholipids. The simplest is phosphatidic acid which is present transiently as an intermediate that is converted to other phospholipids when a second group such as choline, ethanolamine, serine, or glycerol is attached to the phosphate by an enzyme-catalyzed reaction. The fatty acids of phospholipids are typically in the range of 14 to 18 carbons in length. Shorter chains are unable to assemble into stable membranes, while longer chains can potentially "freeze" into a gel state at ordinary temperature ranges. As will be discussed later, all membranes must be in a fluid state in order to function.

Phospholipids typically have one saturated and one unsaturated fatty acid linked to glycerol through ester bonds. Stearic acid, with 18 carbons, is an example of a saturated fatty acid. Oleic acid also has 18 carbons but with a *cis* unsaturated bond between the 9 and 10 carbons. The *cis* double bond puts a kink into the chain (Plate 5.2) that dramatically lowers its melting point because it prevents the chains from packing tightly. For instance, the melting point of stearic acid is 68° C, while oleic acid melts at 14° C. Even more dramatic, a phospholipid with two 18 carbon saturated fatty acids (distearoylphosphatidylcholine) has a phase transition temperature of 55° C but this decreases to −17° C when a single double bond is added to both chains.

Although most phospholipids have fatty acids attached through ester bonds, a few lipids, particularly those of extremophilic microorganisms like the Archaea, use ether linkages which are more stable to hydrolysis. Another accommodation to high temperatures and extreme pH ranges is hydrocarbon chains with several methyl groups attached instead of unsaturated bonds, such as phytic acid (Plate 5.2). Unsaturated bonds are damaged by reacting with molecular oxygen in air, while methyl groups resist oxidation. The branched chains serve the same purpose as unsaturation by increasing the fluidity of the membranes that would otherwise freeze into a gel state.

Self-Assembly of Amphiphiles

All lipids are surface active and would be classified as surfactants that accumulate at the air-water interface as shown in Figure 5.2. Monomolecular layers are the simplest example of self-assembly. The hydrocarbon chain of a surfactant molecule is relatively insoluble in water and therefore referred to as hydrophobic, while the hydrophilic head group interacts strongly with water. The result is that the hydrophobic groups form a continuous oily layer between the water and air and are anchored there by the hydrophilic groups interacting with the water. However, the amphiphilic molecules are not exclusively at the surface, they also are dispersed as solutes in the bulk phase, which assemble into micelles and vesicles called liposomes. Micelles have a purely hydrophobic interior composed of hydrocarbon chains, while liposomes are defined by a lipid bilayer boundary that contains an interior aqueous volume (Fig. 5.2). It is significant that when amphiphilic vesicles are dried they preserve the bilayers, which are now stacked into a multilamellar structure. Furthermore, during the drying process any solutes that are present in the solution become trapped between the head groups of the lipids. *This essential process is the simplest way for protocells containing polymers to assemble in prebiotic conditions.*

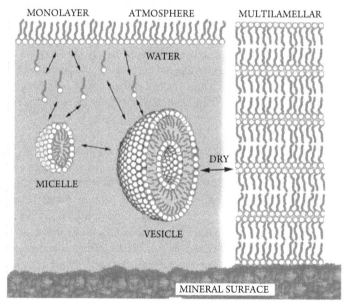

Figure 5.2. Self-assembled structures of amphiphilic molecules in water and after drying. Single-chain amphiphilic molecules such as a monocarboxylic acid are surfactants, meaning that some are in solution while others form a monolayer at the interface between water and the atmosphere. When dried, the vesicles fuse to produce a multilamellar structure. The arrows indicate that there is continuous exchange and transformation among the various phases.

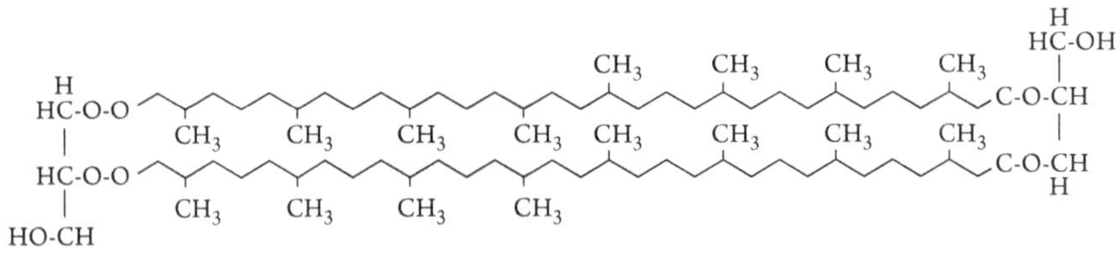

```
                                                                          H
                                                                         HC-OH
                                CH₃      CH₃      CH₃      CH₃             |
   H                                                                     C-O-CH
  HC-O-O                  CH₃    CH₃    CH₃    CH₃  CH₃    CH₃    CH₃  CH₃  |
   |                                                                     C-O-CH
  HC-O-O                  CH₃    CH₃    CH₃    CH₃                         H
   |
  HO-CH
```

Figure 5.3. A characteristic tetraether lipid that forms stable membranes of hyperthermophilic microorganisms. Note that the chemical bonds linking the hydrocarbon chain to glycerol are ethers (-C-O-C-) rather than the ester bonds of phospholipids. Ether bonds are much less labile to hydrolysis.

In the laboratory, it is possible to cause amphiphilic compounds to form what are called planar bilayers on microscopic apertures. For instance, when a phospholipid mixture is spread across a small aperture in a thin sheet of a hydrophobic plastic, it becomes a lipid bilayer supported by the edge of the aperture. Plate 5.3 shows a computer simulation of a planar bilayer in which hydrophilic head groups (blue) interact with water molecules (red), while hydrophobic hydrocarbon chains (light blue) are in the interior. The chains appear disordered because they are in a fluid state. Planar lipid bilayers are important model systems that provide valuable information about membrane permeability and how protein channels control solute transport across cell membranes.

Certain microorganisms called hyperthermophiles thrive in the near-boiling-water temperatures of hot springs. Lipid bilayers are not stable at those temperature ranges, so hyperthermophiles synthesize tetraether lipids, with one example shown in Figure 5.3. The hydrocarbon chains are long enough to span the membrane and form a monolayer in the membrane, rather than a bilayer. The molecules are highly branched with multiple methyl groups ($-CH_3$) along the chains. The hydrophilic ends are simply glycerol rather than phosphate and are attached to the chains by ether linkages rather than ester bonds.

Forces Stabilizing Self-Assembly of Amphiphiles

Why can certain lipids spontaneously assemble into bilayer structures? One stabilizing force is van der Waals interactions, named after the 19th century scientist who first described them. The electron shells of molecules are not fixed around the nuclei of the atoms that compose them, but instead experience fluctuations that cause slight transient differences in the electrical charge of the surface. When two molecules are near each other, the fluctuating charges on one surface induce opposite charges in the second surface, producing an attractive force. Although van der Waals interactions are much weaker than ionic and covalent bonds between atoms, in aggregate the forces can have a significant effect on the physical properties of molecules. For example, methane, ethane, propane, and butane with 1, 2, 3, and 4 carbons are all gases at ordinary temperatures and pressure, but pentane, with 5 carbons, has sufficient van der Waals interactions among molecules to be a liquid. All saturated hydrocarbons up to hexadecane with 16 carbons are liquids at room temperature; longer hydrocarbons such as those composing paraffin are solid.

The second physical effect that stabilizes lipid bilayers emerges from the way the hydrocarbon chains interact with water. If a single hydrocarbon molecule like decane, with ten carbon atoms, could somehow be forced into water, the presence of the chain would cause hydrogen bonds between water molecules to dissociate, so that energy is added to the system. The presence of the chain causes water molecules to form an ordered layer of water around the chain, and the ordering effect also requires energy that is stored in the system as free energy.

How can the energy be released? If one chain happens to encounter another chain, bringing the two chains together allows hydrogen bonds in the water to form again, and also allows the organized water molecules to become disordered. A principle of thermodynamics is that if a system can become more disordered, it will. In other words, the entropy of the system increases if hydrocarbon chains are together rather than being dissolved in the water. This process is sometimes called the hydrophobic effect, or entropic bonding, and after a while, virtually all of the decane molecules and water will become two separate phases, fulfilling the old saying that water and oil don't mix. The hydrocarbon chains of fatty acids and phospholipids are also subject to the hydrophobic effect, so they spend most of the time interacting with other hydrocarbon chains in lipid bilayers.

Membrane Fluidity

In the 1960s, it began to be understood that the proteins of biological membranes are embedded in a fluid sea of lipids. The first indication that membranes are surprisingly fluid was an experiment by Frye and Ededin (1970) who used fluorescent dyes to label the membranes of cells, then caused the cells to fuse. They observed that the labeled components of two separate membranes began to diffuse into each other within minutes and concluded that the membrane had a fluid character rather than being fixed in place. The fluidity of lipid bilayers has been clearly demonstrated by using a focused laser light beam to bleach a small area of a fluorescent lipid bilayer (Fig. 5.4). Within seconds, the diffusion of lipids within the bilayer began to fill in the bleached spot, which completely disappeared five minutes later.

Numerous studies have measured the diffusion coefficient of lipids and proteins in membranes, and the diffusion rates were found to correspond to those expected of a fluid with the viscosity of oil, rather than a gel phase resembling wax. The fluidity of the lipid phase has ramifications for the origin of life because a stable prebiotic membrane

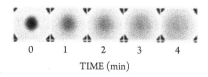

TIME (min)

Figure 5.4. Demonstration of membrane fluidity in a lipid bilayer. Adapted from Cho et al. 2013.

not only requires a certain hydrocarbon chain length of 10 or more carbons, but also the physical property of fluidity.

Prebiotic Self-Assembly of Membranes

Given that mixtures of amphiphilic compounds can be synthesized by abiotic reactions, as described in Chapter 4, is it plausible that such simple compounds could assemble into stable membranes in the prebiotic environment? In fact, amphiphilic compounds extracted from the Murchison meteorite can assemble into membranous vesicles that have significant stability (Deamer, 1985; Deamer and Pashley, 1989). Plate 5.4 shows two micrographs illustrating vesicle formation when a dried extract of the meteorite interacts with an aqueous phase. The vesicles are presumably composed of a mixture of monocarboxylic acids that are known to be present in the meteorite. The vesicles are also highly fluorescent, indicating that they contain polycyclic aromatic hydrocarbons (PAH) such as pyrene and fluoranthene.

The organic compounds composing the membranes probably predated the formation of the solar system, since the asteroid parent body of the Murchison meteorite formed from interstellar dust present in the molecular cloud from which the sun and planets emerged 4.5 Gya. It was encouraging to observe that such ancient organic compounds can assemble into vesicles, but there is still much to learn about the assembly and composition of prebiotic membranes.

Primitive Membrane Functions: Permeability

The simplest function of the lipid bilayer in a biological membrane is to provide a permeability barrier limiting free diffusion of ionic and polar solutes through the membrane. Although such barriers are essential for cellular life to exist, there must also be a mechanism by which selective permeation allows specific solutes to cross the membrane. In contemporary cells, transport of ions and nutrients is mediated by transmembrane proteins that act as channels and transporters. Examples include the proteins that facilitate the transport of glucose and amino acids into the cell, channels that allow potassium and sodium ions to permeate the membrane, and active transport of ions by enzymes that use ATP as an energy source.

To give a perspective on permeability and transport rates by diffusion, we can compare the fluxes of relatively permeable and relatively impermeable solutes through lipid bilayers. A quantitative measure of permeability is given by a parameter called the permeability coefficient (P), which is determined by measuring the flux of a solute across a unit area of membrane driven by a concentration gradient of the solute. A simple version of the equation defining the permeability coefficient is $P = J/\Delta C$, where J is the measured flux expressed as moles of solute $cm^{-2}\,s^{-1}$, and ΔC is the difference in concentration of the solute across the membrane with units of moles cm^{-3}. The units of P are expressed as centimeters per second ($cm\,s^{-1}$). Although the measured values of permeability coefficients vary considerably depending on the lipid composition of the bilayers, a typical permeability coefficient of water diffusing through lipid bilayer membranes is approximately 10^{-4} cm s^{-1}, while potassium ions cross the bilayer at much slower rates with a measured permeability coefficient of 10^{-12} cm s^{-1}. It follows that the measured permeability of lipid bilayers to small, uncharged molecules such as water is greater than the permeability to ions by a factor of ~10^8. These values mean little by themselves but make more sense when put in the context of time required for exchange across a bilayer: half the water in a liposome exchanges in a few milliseconds, while potassium ions have half-times of exchange measured in days.

The reason that potassium ions are so impermeable arises from an effect called Born energy, named after Max Born who pointed out that ions have a property associated with their electric field which he called self-energy. As a result of the field, ions in a polar medium like water strongly interact with the electrical charges on the water molecules, and those interactions must be broken to move the ion into a nonpolar medium like the oily phase of a lipid bilayer. This requires a very large expenditure of energy, one estimate being ~40 kcal mole^{-1}, so virtually no ions would be able to penetrate the bilayer. The fact that they do, albeit at a very slow rate, can be accounted for by transient fluctuations in the bilayer structure that allow hydrated ions to cross the barrier (Paula et al., 1996).

The flux of water and ions across lipid bilayers are examples of the most permeable and least permeable solutes; most other solutes such as metabolites and sugars fall between the two extremes. Glucose, for instance, has a permeability coefficient of 5×10^{-11}, 50-fold greater than typical monovalent cations. Significantly, ribose, with just one less carbon in its ring, has a surprisingly high permeability of 2×10^{-7}, approximately 4000 times faster than glucose (Sacerdote and Szostak, 2005) and ten times faster than other 5-carbon sugars. On the other hand, zwitterionic amino acids such as glycine and anions such as phosphate exhibit very low permeability in the same range as monovalent cations (Chakrabarti and Deamer, 1992).

The extreme permeability barrier to ions means that ionic concentration gradients produced by active transport can be maintained across the boundary membranes of cells or subcellular compartments like mitochondria and chloroplasts. This property, combined with ion specific protein channels and active ion transport processes allows living cells to maintain and regulate the composition of an internal volume that is very different from the external aqueous medium.

Insertion of Peptides and Proteins into Lipid Bilayers

It is important, in a book about the origin of life, to note the astonishing complexity of membrane functions because each function evolved from simpler predecessors in primitive life, and the nature of the predecessor is a vast gap in our knowledge of how life began. The fluid-mosaic membrane structure was a revelation when it was proposed in 1972, and over the following 40 years the model clarified not just the structure of biological membranes but also their functions. The term "mosaic" comes from the idea that proteins embedded in a lipid bilayer are like tiles embedded in the plaster of a mosaic work of art. The simplest function of an embedded protein is to provide a hydrated channel through the bilayer so that certain ions like potassium and sodium can cross the bilayer barrier. Some ion channels have evolved into gates that open only when activated by voltage or by the binding of a ligand like acetylcholine. Other proteins evolved into specific carriers that allow the cell to access essential nutrients like amino acids and carbohydrates by transporting them across the bilayer barrier.

The electron transport enzymes in the membranes of aerobic bacteria and their mitochondrial descendants in eukaryotic cells are a more complex set of embedded proteins. These catalyze reactions in which a reduced substrate such as NADH (nicotinamide adenine dinucleotide hydride) donates electrons to a dehydrogenase and the electrons are transferred through a chain of enzymes, ending up on molecular oxygen. Electron transport enzymes also function in photosynthesis, except that the source of electrons is a photochemical reaction in which light energy excites a pigment system that strips electrons from water then transfers them to an acceptor, NADP (nicotinamide adenine dinucleotide phosphate). In both cases, the electron transport is coupled to proton transport so that an electrical potential or pH gradient is produced and maintained across the membrane. The proton gradient is the energy source driving one of the more complex membrane-associated proteins, the ATP (adenosine triphosphate) synthase that can either pump protons using the energy of ATP hydrolysis or be driven in reverse so that protons of a gradient pass through the pumping mechanism and synthesize ATP, the energy currency of all life.

Probably the most complex membrane-associated proteins are those of the photosystems that capture light energy in the cyanobacteria that ultimately evolved into the chloroplasts of plant life today.

Because the insertion of proteins in membranes is an essential first step allowing them to function, we can begin by asking how a water-soluble protein can spontaneously become embedded in a lipid bilayer. A clue to answering this question is the fact that surprisingly simple peptides easily perform this trick. Hladky and Hayden (1972) found that an antibiotic called gramicidin could assemble within a lipid bilayer to form cation-selective channels that allowed sodium and potassium ions to diffuse across the membrane (Fig. 5.5). Their discovery showed that it was possible to experimentally model the ion-selective channels that are present in the membranes of all living cells. Over the next decade, it turned out that gramicidin was one of many antibiotic molecules produced by microorganisms that could alter the permeability of lipid bilayers to ions. Other examples include valinomycin, nigericin, and alamethicin, all of which are small peptides. Gramicidin, for instance, has just 15 amino acids that turn into a partial helical barrel in the hydrophobic interior of a lipid bilayer. A single gramicidin is not large enough to be a channel by itself, but when two gramicidin molecules happen to meet, they form a transient head-to-head hydrogen bond lasting 1–2 seconds. This briefly stabilizes their combined helical structures which span the bilayer in the form of a tunnel-like barrel that accommodates a strand of ~6 water molecules. Ions like sodium or potassium can pass through the tunnel, pushing the water molecules ahead of them.

Certain proteins can also assemble into much larger transmembrane channels that conduct ions. In 1970, an antibiotic protein was reported to be produced by *Staphylococcus* bacteria and named alpha-hemolysin (Sengers, 1970). Over the ensuing years, interest in the pore-forming ability of alpha-hemolysin slowly increased, and in 1987 it was demonstrated to form pores in lipid bilayers (Belmonte et al., 1987). This explained its cytotoxic effect and in particular why it causes erythrocytes to swell and release hemoglobin, a process called hemolysis from which hemolysin gets its name.

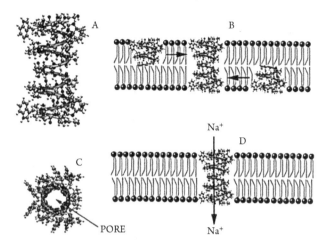

Figure 5.5. Insertion of a peptide into a lipid bilayer. The antibiotic peptide called gramicidin is composed of 15 amino acids and can spontaneously insert partway through a lipid bilayer. When two gramicidin molecules meet, they transiently connect to produce an ion-conducting channel through the bilayer. A. connected pair of gramicidin peptides; B. connection being made in a lipid bilayer; C. lengthwise view of the pore; D. one sodium ion is shown moving through the pore.

A general rule is that selective permeability to ions in the membranes of living cells is provided by specialized transmembrane proteins that produce a water-filled channel through the hydrocarbon layer.

The process by which proteins become embedded in membranes is also significant for understanding how the earliest cells could have access to nutrients present as solutes in the local environment. For instance, something as simple as phosphate is an essential component of metabolism and nucleic acids today and must have been involved in the primitive metabolic chemistry of the first cells. The problem is that lipid bilayers have very low permeability to phosphate, so how could it be available for intracellular metabolism? If peptides were among the polymers being synthesized by prebiotic condensation reactions, some of them were likely to have properties that allowed them to insert into bilayers, just as gramicidin does in the example just presented. The peptides would produce ion-conducting channels through the bilayers that allow phosphate and other nutrients like amino acids to enter the cell.

Encapsulation of Polymers

As described earlier, the self-assembly properties of phospholipids can be visualized simply by drying a lipid solution to produce a thin film, then adding water or a salt solution and shaking the container to disperse the lipids. This results in what are now referred to as multilamellar vesicles, or MLVs. Most of the lipid is concealed in the MLVs but can be dispersed by exposing them to sonication in which high frequency sound disrupts the multilamellar structure and produces small unilamellar vesicles, or SUVs. These are bounded by a single lipid bilayer, but they are also so small that only a tiny fraction of the solution in which they are being made is encapsulated. For instance, a typical SUV might be just 20 nanometers (nm) in diameter. The bilayer membrane is 5 nm thick, so the internal volume has a diameter of 10 nm. If the aim is to capture a solution of 1 mM ATP in the vesicles, simple arithmetic shows that each SUV would only contain an average of two ATP molecules!

Calculations like this led to attempts to produce larger vesicles, and several methods were soon established. In one method, phospholipids were dissolved in a detergent solution followed by dialysis to remove the detergent. The lipid molecules assembled into a reasonably homogeneous preparation of vesicles with diameters in the micrometer range. However, the simplest method still in use is to pass an MLV preparation through a filter having a known pore size. As the MLVs are forced through the pores under pressure, they break up into smaller vesicles with fewer lipid layers. After ten passages most of the lipid vesicles are unilamellar and have approximately the same diameter as the pores.

Variations of the above methods are used in industry to produce liposomes containing pharmaceutical agents, but they are highly technical and could not work under prebiotic conditions. Is there a plausible prebiotic process? My students and I addressed this question in the early 1980s and found an answer that led to the ideas presented in this book. In early research using freeze-fracture electron microscopy to visualize the structures produced by self-assembling lipids, it became clear that lipid vesicles in water fused into multilamellar structures when dried. Furthermore, anything dissolved in the water would be trapped between the lamellae composed of lipid bilayers. Upon rehydration, vesicles would bud off the dry lipid as water penetrated between bilayers. Half the vesicles would be empty because they were produced by fusion of vesicle interiors, but the other half would contain the solute because they happened to fuse so that their external surfaces became internal surfaces. The method is a simulation of the wet–dry cycle that would occur on mineral surfaces in the prebiotic environment (Deamer and Barchfeld, 1982; Shew and Deamer, 1985).

The basic process illustrated in Figure 5.6 shows small lipid vesicles dispersed in an aqueous phase undergoing fusion as water evaporates. Note that every other layer composed of lipid head groups is empty and corresponds to the original interior of the vesicles. The alternate layers are filled with highly concentrated solute molecules which are trapped between external surfaces when they fuse (A and B). The image labeled C shows a freeze-fracture micrograph of multilamellar phospholipids, and D shows fluorescently labeled nucleic acids encapsulated in lipid vesicles.

Although phospholipids were used in these experiments, they are unlikely to serve as prebiotic amphiphiles because they are synthesized by complex enzyme-catalyzed reactions. Therefore, it was necessary to demonstrate that more plausible amphiphiles can also capture macromolecules by a wet–dry–wet cycle. The vesicles are not composed of a phospholipid but instead of a much simpler amphiphile called lauric acid with 12 carbons in its chain. The lauric acid was mixed with its monoglyceride, and the mixture readily formed the vesicles shown in Figure 5.7A. When the mixture was dried in the presence of rRNA from yeast, the RNA was encapsulated (Fig. 5.7B). This is by far the simplest way to produce protocells and could occur in hydrothermal freshwater

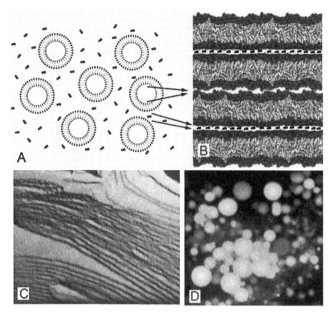

Figure 5.6. Illustration of fusion and encapsulation. The small unilamellar vesicles (A) fuse into a multilamellar structure (B) when dried, capturing solutes between the head groups of alternate lipid layers. The structure of multilamellar lipids is revealed by freeze-fracture electron microscopy (C) which shows each bilayer as a dark line 5 nm thick. When the multilamellar structure is rehydrated by adding water, vesicles reform and encapsulate up to half of the original solute, such as the fluorescent DNA shown in (D). The vesicles range from 10 to 30 micrometers in diameter.

on the prebiotic Earth. In fact, the process illustrated in the figure was carried out with water taken from a hot spring in Yellowstone National Park (Milshteyn et al., 2018). Joshi et al. (2017) have reported similar self-assembly properties of fatty acids in the water of a hydrothermal site in India.

Conclusions and Open Questions

It is reasonable to assume that hydrocarbons and their amphiphilic derivatives were available in the prebiotic environment. In freshwater conditions, amphiphiles assemble spontaneously into membranous compartments bounded by lipid bilayers if chain lengths are 10 or more carbons long. The compartments readily encapsulate solutes, so if polymers were synthesized by a second process, the result would be protocells containing chance assemblies of polymers having random sequences. Each protocell is a natural experiment, a kind of test to determine whether a particular system of polymers has functional properties related to stability, pore formation, growth and replication.

Significantly, the presence of divalent cations like magnesium and calcium strongly inhibits membrane assembly from fatty acids. This means that a pure fatty acid

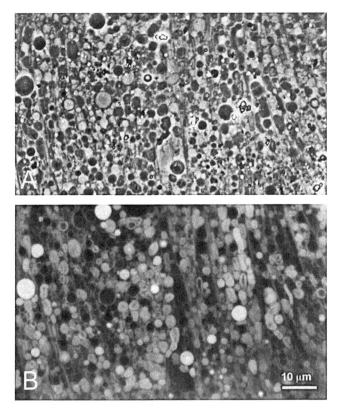

Figure 5.7. Membranous vesicles spontaneously form (A) when a mixture of fatty acid and its monoglyceride is dispersed in water from a hot spring in Yellowstone National Park (Milshteyn et al., 2018). If the same mixture is exposed to a single evaporation–rehydration cycle in the presence of short strands of RNA stained with acridine orange, a fluorescent dye, the concentrated polymer is encapsulated within membrane-bounded vesicles (B).

membrane would be unstable in marine environments due to the high concentrations of Mg^{2+} (53 mM) and Ca^{2+} (10 mM). This concern was first addressed by Monnard et al. (2002). Fatty acid membranes have been employed extensively by the Szostak research group as model membranes of primitive life (Hanczyc et al., 2003; Hanczyc et al., 2007). Magnesium ions disrupted the membranes, but Adamala and Szostak (2013) reported that addition of citrate to the mixture chelated the Mg^{2+} so that it did not damage the membranes but could still be available as an essential cofactor for RNA chemistry. Acidic pH ranges represent another limitation for fatty acid membranes, because the neutral protonated forms cannot assemble into stable membranes. Namani and Deamer (2008) reported that addition of a long-chain amine overcame this limitation by adding a positive charge at low pH ranges.

These observations put constraints on the composition of membranes that could assemble in the prebiotic environment. Although monocarboxylic acid chain lengths detected in the Murchison meteorite range from 8 to 12 carbons in length, the longer chain lengths are present only at a few ppm (Naraoka et al., 1999). An open question, then, concerns a source of amphiphilic compounds having chain lengths sufficient for stable bilayers.

Another problem is that membranes assembled from pure single chain amphiphiles are relatively fragile compared to membranes formed by phospholipids having two chains. But, as Budin et al. (2014) observed, membranes containing fatty acid mixtures are more stable than those formed by a pure compound—one way that additional stability could be imparted to prebiotic membranes. Furthermore, Groen et al. (2012) found that small amounts of PAHs such as pyrene have a stabilizing effect resembling that of cholesterol in the membranes of cells today. In this regard, Black and Blosser (2016) reported the surprising result that the nucleobase adenine can stabilize fragile fatty acid membranes against the disruptive effects of salt.

We conclude that the most plausible conditions under which membranes could be incorporated into the earliest life would be in moderately acidic (pH ~3-5) freshwater ponds in which the concentration of divalent cations is low enough so that they do not interfere with the assembly process. The membrane composition would not be a pure compound, but instead would be composed of mixtures of amphiphilic compounds with chains of 10 carbons or longer mixed with other nonpolar compounds such as PAH to promote stability.

References

Adamala K, Szostak JW (2013) Competition between model protocells driven by an encapsulated catalyst. *Nat Chem* 5, 495–501.

Bangham AD, Standish MM, Watkins JC (1965) Diffusion of univalent ions across the lamellae of swollen phospholipids. *J Mol Biol* 13, 238–252.

Belmonte G, Cescatti L, Ferrari B, Nicolussi T, Ropele M, Menestrina G (1987) Pore formation by *Staphylococcus aureus* alpha-toxin in lipid bilayers: dependence upon temperature and toxin concentration. *Eur Biophys J* 14, 349–358.

Black RA, Blosser MC (2016) A self-assembled aggregate composed of a fatty acid membrane and the building blocks of biological polymers provides a first step in the emergence of protocells. *Life* 6, 33. doi:10.3390/life6030033.

Budin I, Prywes N, Zhang N, Szostak JW (2014) Chain-length heterogeneity allows for the assembly of fatty acid vesicles in dilute solutions. *Biophys J* 107, 1582–1590.

Chakrabarti A, Deamer DW (1992) Permeability of lipid bilayers to amino acids and phosphate. *Biochim Biophys Acta* 1111, 171–177.

Chen IA, Szostak JW (2004) A kinetic study of the growth of fatty acid vesicles. *Biophys J* 87, 988–998.

Cho N-J, Hwang LY, Solandt JJR, Frank CW (2013) Comparison of extruded and sonicated vesicles for planar bilayer self-assembly. *Materials* 6, 3294–3308.

Danielli JF, Davson H (1935) A contribution to the theory of permeability of thin films. *J Cell Comp Physiol* 5, 495–500.

Deamer DW (1985) Boundary structures are formed by organic compounds of the Murchison carbonaceous chondrite. *Nature* 317, 792–794.

Deamer DW, Barchfeld GL (1982) Encapsulation of macromolecules by lipid vesicles under simulated prebiotic conditions. *J Mol Evol* 18, 203–206.

Deamer DW Pashley R (1989) Amphiphilic components of the Murchison carbonaceous chondrite: Surface properties and membrane formation. *Orig Life Evol Biosph* 19, 21–38.

Dyson F (1999) *Origins of Life.* Cambridge University Press, Cambridge UK

Fraccia TP, Smith GP, Zanchetta G, Paraboschi E, Youngwoo Y, Walba DM, Dieci G, Clark NA, Bellini T (2015) Abiotic ligation of DNA oligomers templated by their liquid crystal ordering. *Nat Comm* 6. doi:10.1038/ncomms7424.

Frye LD, Edidin M (1970) The rapid intermixing of cell surface antigens after formation of mouse-human heterokaryons. *J Cell Sci* 7, 319–335;

Groen J, Deamer D, Kros A, Ehrenfreund P (2012) Polycyclic aromatic hydrocarbons as plausible prebiotic membrane components. *Orig Life Evol Biosph* 42, 295–306.

Gorter E, Grendel F (1925) On biomolecular layers of lipoids on the chromocytes of the blood. *J Exp Med* 41, 439–443.

Hanczyc MM, Fujikawa SM, Szostak JW (2003) Experimental models of primitive cellular compartments: Encapsulation, growth, and division. *Science* 302, 618–622.

Hanczyc MM, Mansy S, Szostak JW (2007) Mineral surface directed membrane assembly. *Orig Life Evol Biospheres* 37, 67–82.

Hargreaves WR, Deamer DW (1978) Liposomes from ionic, single-chain amphiphiles. *Biochemistry* 17, 3759–3768.

Hladky SB, Hayden DA (1972) Ion transfer across lipid membranes in the presence of gramicidin A: I. Studies of the unit conductance channel. *Biochim Biophys Acta—Biomembranes* 274, 294–312.

Joshi MP, Samanta A, Tripathy GR, Rajamani S (2017) Formation and stability of prebiotically relevant vesicular systems in terrestrial geothermal environments. *Life* 7, 51. doi:10.3390/life7040051.

Mansy SS, Szostak JW (2008) Thermostability of model protocell membranes. *Proc Natl Acad Sci USA* 105, 13351–13355.

Mendoza N (1995) Alfonso L. Herrera: A Mexican pioneer in the study of chemical evolution. *J Biol Phys* 20, 11–15.

Milshteyn D, Damer B, Havig J, Deamer D (2018) Amphiphilic compounds assemble into membranous vesicles in hydrothermal hot spring water but not in seawater. *Life (Basel)* doi: 10.3390/life8020011

Monnard, P-A, Apel CL, Kanavarioti A, Deamer DW (2002) Influence of ionic inorganic solutes on self-assembly and polymerization processes related to early forms of life: Implications for a prebiotic aqueous medium. *Astrobiology* 2, 139–152.

Namani T, Deamer DW (2008) Stability of model membranes in extreme environments. *Orig Life Evol Biosph* 38, 329–341.

Naraoka H, Shimoyama A, Harada K (1999) Molecular distribution of monocarboxylic acids in Asuka carbonaceous chondrites from Antarctica. *Orig Life Evol Biospheres* 29, 187–201.

Overton E (1895) Über die osmotischen Eigenschaften der lebenden Pflanzen- und Tierzellen. Vierteljahresschr. *Naturforsch Ges Zürich* 40, 159–201.

Paula S, Volkov AG, van Hoek AN, Haines TH, Deamer DW (1996) Permeation of protons, potassium ions, and small polar molecules through phospholipid bilayers as a function of membrane thickness. *Biophys. J.* 70, 339–348.

Robertson JD (1967) Origin of the unit membrane concept. *Protoplasma* 63, 218–245.

Sacerdote MG, Szostak JW (2005) Semipermeable lipid bilayers exhibit diastereoselectivity favoring ribose. *Proc Natl Acad Sci USA* 102, 218–245.

Sengers RCA (1970) Hemolytic action of staphylococcal α-hemolysin on human erythrocytes in a Na+- and K+-containing suspending fluid. *Antonie van Leeuwenhoek* 36, 57–65.

Shew RL, Deamer DW (1985) A novel method for encapsulation of macromolecules in liposomes. *Biochim Biophys Acta—Biomembranes* 816, 1–8.

Singer SJ, Nicolson GL (1972) The fluid mosaic model of the structure of cell membranes. *Science* 175, 720–731.

Tanford C (1973) *The Hydrophobic Effect: Formation of Micelles and Biological Membranes.* New York: Wiley.

Walde P, Wick R, Fresta M, Mangone A, Luisi PL (1994) Autopoietic self-reproduction of fatty acid vesicles. *J Am Chem Soc* 116, 11649–11654.

Yakovchuk P, Protozanova E, Frank-Kamenetskii MD (2006) Base-stacking and base-pairing contributions into thermal stability of the DNA double helix. *Nucleic Acids Res.* 34, 564–574.

Condensation Reactions Synthesize Random Polymers

CO-AUTHORED WITH DAVID ROSS

It has not escaped our notice that the specific pairing that we have postulated immediately suggests a possible copying mechanism for the genetic material.

James Watson and Francis Crick, 1953.

Overview and Questions to Be Addressed

Over the past half century of serious research on the origin of life, several schools of thought have emerged that focus on "worlds" and what came first in the pathway to the origin of life. One example is the RNA World, a term coined by Walter Gilbert after the discovery of ribozymes. Other examples include the Iron-Sulfur World of Günther Wächtershäuser and the Lipid World proposed by Doron Lancet and coworkers. Then we have a competition between "metabolism first" and "replication first" schools. The worlds and schools have the positive effect of sharpening arguments and forcing us to think carefully, but they also can lock researchers into defending their individual approaches rather than looking for patterns in a larger perspective. One of the main themes of this book is the notion that the first living cells were systems of functional polymers working together within membranous compartments. Therefore, it is best not to think of "worlds" and "firsts" as fundamentals but instead as components evolving together toward the assembly of an encapsulated system of functional polymers. At first the polymers will be composed of random sequences of their monomers, and the compartments will contain random assortments of polymers. Here, we refer to these structures as protocells which are being produced in vast numbers as they form and decompose in continuous cycles driven by a variety of impinging, free-energy sources.

This chapter describes how thermodynamic principles can be used to test the feasibility of a proposed mechanism by which random polymers can be synthesized. There is a current consensus that early life may have passed through a phase in which RNA served as a ribozyme catalyst, as a replicating system, and as a means for storing and expressing genetic information. For this reason, we will use RNA as a model polymer, but condensation reactions also produce peptide bonds and oligopeptides. At some point in

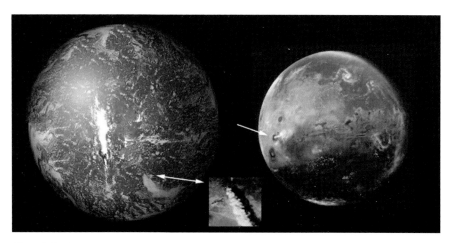

Plate 1.1. The early Earth and Mars were both supplied with water that formed a global ocean on Earth and shallow seas on Mars. Volcanoes, indicated by arrows, emerged on both planets. *Source*: Assembled by author from public domain images.

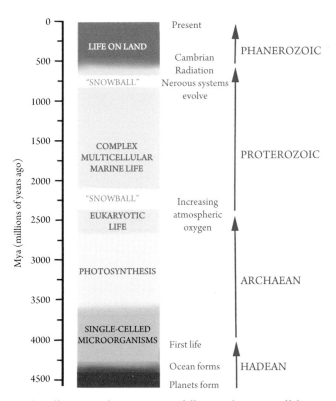

Plate 1.2. Timeline illustrating the main events following the origin of life on Earth, during four eras: Hadean, Archean, Proterozoic, and Phanerozoic. The words are derived from Greek words: *Hadean* from Hades, the god of the underworld because of the elevated temperatures in that era, and *Archean* from archea meaning ancient. *Proterozoic* combines protero- (earlier) and zoic- (animal), and *Phanerozoic* combines phanerós (visible) and zoic, meaning visible life, since it was once believed that life began in the Cambrian about half a billion years ago. Note also the abbreviation on the left: Mya for millions of years ago. We will use this abbreviation in many of the chapters to follow, and also Gya for billions of years ago, again derived from Greek words: mega- from a word meaning large and giga- from a word meaning giant. There is good evidence that the Earth was nearly covered in ice or an icy slush at least twice; these periods are referred to as "snowball Earths."

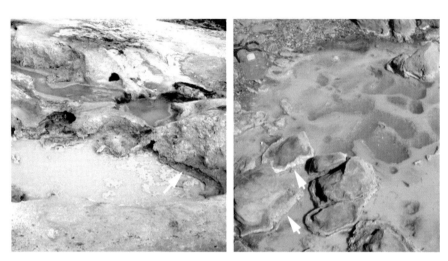

Plate 2.1. Clay-lined pools in hydrothermal fields on Mount Lassen (left) and Mount Mutnovsky (right). The typical pH of such pools is acidic, pH 2 to 4, and temperatures range from 70 to over 90° C. Arrows show areas where evaporation has left films of concentrated solutes. Photographs by author.

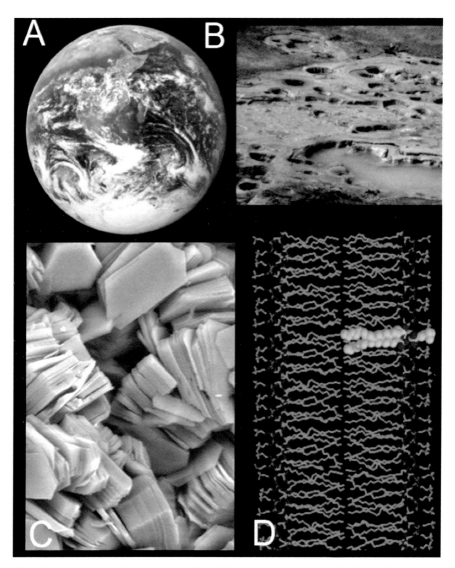

Plate 2.2. The four scales illustrated: The global scale of the Earth (A), the local scale of a hydrothermal field (B), the microscopic scale of clay particles (C), and the nanoscale of a lipid bilayer membrane (D).

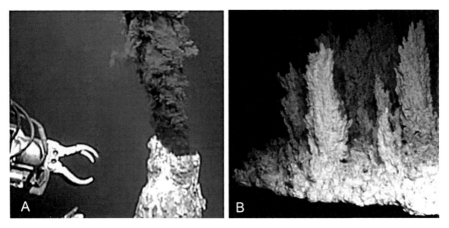

Plate 3.1 A hydrothermal vent emitting "smoke" composed of iron-rich mineral particles (A). The white carbonate mineral structures of an alkaline vent (B). Adapted from public domain images.

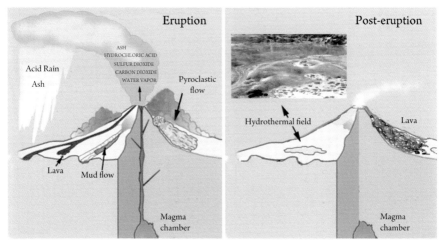

Plate 3.2. Following volcanic eruptions lasting days to weeks, water accumulates in hydrothermal fields that remain for thousands of years.

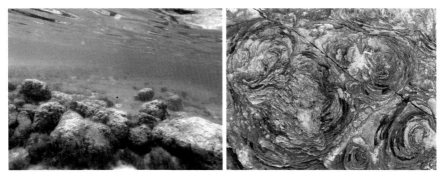

Plate 3.3. Living and fossil stromatolites in the Pilbara region of Western Australia. Left-hand panel shows stromatolites forming in Hamlin Pool, Shark Bay, and the panel on the right shows ancient stromatolite fossils from the Tumbiana region of Pilbara, dated to 2.7 Gya. Credit: Bruce Damer.

Plate 4.1. A molecular cloud in the Carina star-forming region of our galaxy. New stars light up the interior of the cloud; the characteristic jets flowing from the poles of a young rotating star can be seen in the upper right. Credit: NASA Hubble image

Plate 4.2. Overview of solar system formation showing accretion of planetesimals, comets, and dust particles that deliver organic compounds to planetary surfaces. Image adapted from public domain.

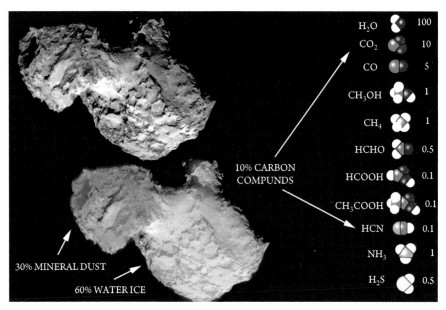

H₂O   100
CO₂   10
CO   5
CH₃OH   1
CH₄   1
HCHO   0.5
HCOOH   0.1
CH₃COOH   0.1
HCN   0.1
NH₃   1
H₂S   0.5

10% CARBON COMPUNDS

30% MINERAL DUST

60% WATER ICE

Plate 4.3. Composition of a typical comet. The amounts of the various carbon, oxygen, nitrogen, and sulfur compounds are shown in relation to water which is taken as 100. The colors on the comet are not real but instead are intended as a visual illustration of the amounts of dust, ice, and organic substances composing a typical comet nucleus. Adapted by author from ESA Rosetta mission image of comet 67P/Churyumov-Gerasimenko.

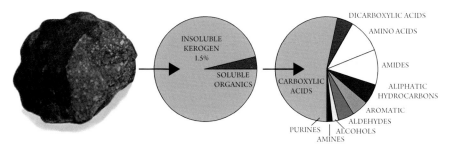

INSOLUBLE KEROGEN 1.5%

SOLUBLE ORGANICS

CARBOXYLIC ACIDS

DICARBOXYLIC ACIDS

AMINO ACIDS

AMIDES

ALIPHATIC HYDROCARBONS

AROMATIC

ALDEHYDES

PURINES   ALCOHOLS

AMINES

Plate 4.4. Organic compounds in the Murchison carbonaceous meteorite. Illustration by author.

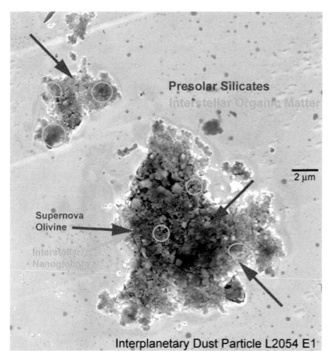

Presolar Silicates

Interstellar Organic Matter

2 µm

Supernova
Olivine

Interstellar
Nanoglobular

Interplanetary Dust Particle L2054 E1

Plate 4.5. Mineral particle collected from a cometary dust stream in 2003 when the Earth passed through the tail of comet 26P/Grigg-Skjellerup. The particle and the organic material on its surface are likely to be left over from the molecular dust cloud from which the solar system formed. Credit: H. Busemann.

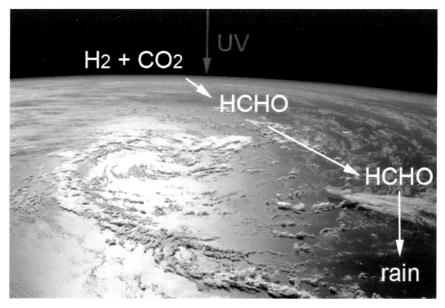

$H_2 + CO_2$

UV

HCHO

HCHO

rain

Plate 4.6 Photochemical synthesis of formaldehyde (CH_2O). Adapted from public domain image of Earth.

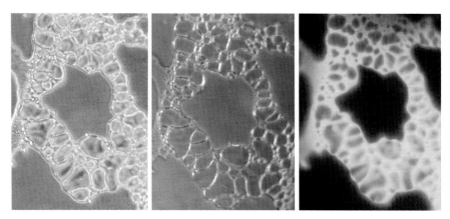

Plate 4.7. Images showing the transformation of pyruvic acid, a relatively simple organic molecule, by pressure and elevated temperature into more complex products with physical properties of self-organization. Micrograph by author in Hazen and Deamer, 2008.

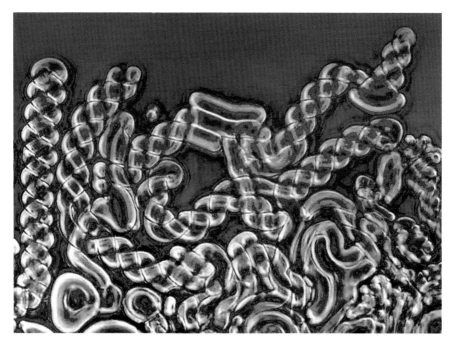

Plate 5.1. Self-assembled tubular structures are produced when dry lecithin is exposed to water. The "tubules" are actually composed of hundreds of concentric lipid bilayers. They are unstable and, given enough time, break up into multilamellar vesicles. The lipids are colorless, so color was added computationally for illustrative purposes. Bar shows 20 micrometers. Photograph by the author.

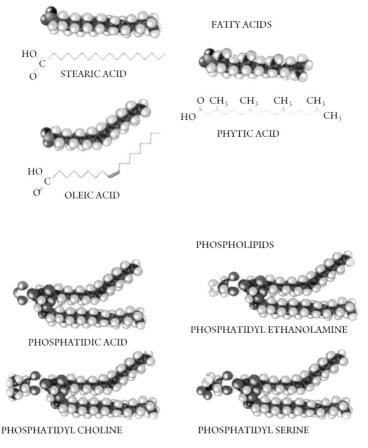

FATTY ACIDS

STEARIC ACID

PHYTIC ACID

OLEIC ACID

PHOSPHOLIPIDS

PHOSPHATIDIC ACID

PHOSPHATIDYL ETHANOLAMINE

PHOSPHATIDYL CHOLINE

PHOSPHATIDYL SERINE

Plate 5.2. Fatty acids are composed of a hydrocarbon chain with a carboxyl group at one end. Some fatty acids are referred to as being saturated, meaning that their chains have no double bonds, while others are unsaturated with one or more *cis* double bonds in the hydrocarbon chain. Stearic acid (saturated) and oleic acid (unsaturated) are shown here, each with 18 carbon atoms. Phytic acid is also illustrated, but instead of unsaturation, methyl groups are present. The phospholipid phosphatidic acid has two fatty acids attached to a glycerol by ester bonds; more complex phospholipids have an additional group linked to the glycerol with an ester bond. Illustration by author.

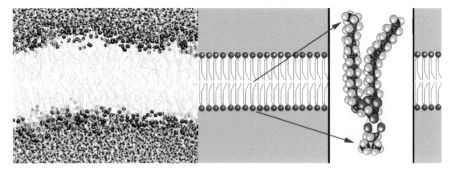

Plate 5.3. Computer simulation of a phospholipid bilayer in water and interpretation showing a phospholipid molecule in a schematic bilayer. The image shows fluid hydrocarbon chains in light blue, lipid head groups in dark blue, and water molecules on either side in red and white. Image courtesy Andrew Pohorille.

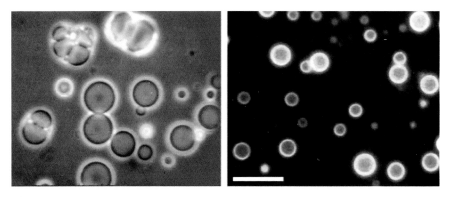

Plate 5.4. Lipid-like amphiphilic material extracted from the Murchison meteorite was separated by thin layer chromatography, and a small amount was dried on a microscope slide, followed by addition of a dilute alkaline buffer. The amphiphilic compounds assembled into vesicles ranging from 10 to 20 micrometers in diameter. The vesicles were fluorescent (panel on right) due to polycyclic aromatic compounds in the mixture. Bar is 20 micrometers. Photographs by author.

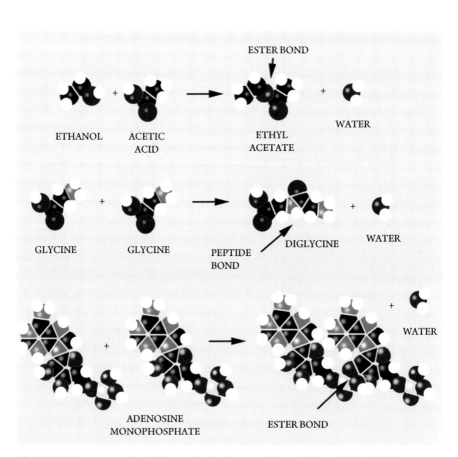

ESTER BOND

ETHANOL   ACETIC
            ACID

ETHYL
ACETATE

WATER

GLYCINE   GLYCINE

PEPTIDE
BOND

DIGLYCINE

WATER

ADENOSINE
MONOPHOSPHATE

ESTER BOND

WATER

Plate 6.1. Three examples of the synthesis of ester and peptide bonds that link the monomers of nucleic acids and proteins: Fischer esterification; synthesis of a peptide bond; and the linking of two mononucleotides with a phosphoester bond. Image prepared by author.

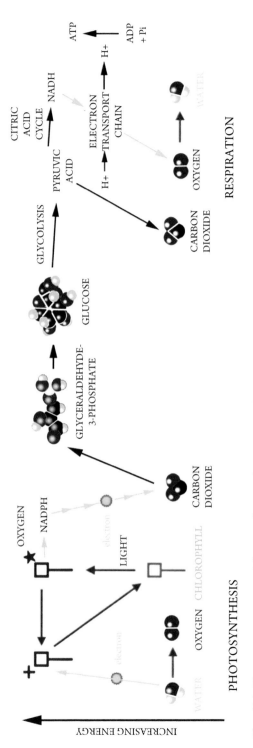

Plate 7.1 The primary reactions of photosynthesis and respiration. Illustration by author.

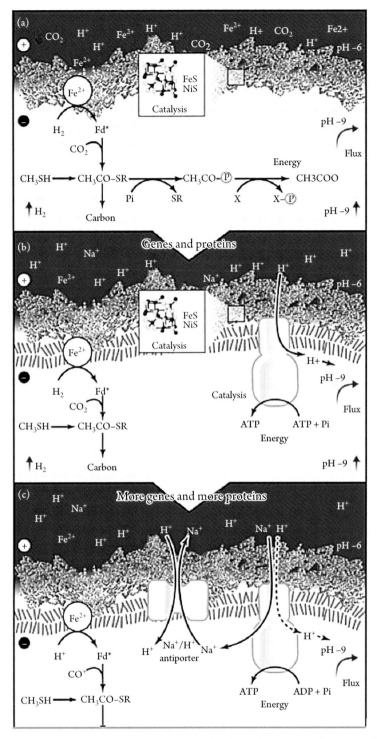

Plate 7.2. Evolution of metabolic reactions in hydrothermal vent minerals, proposed by Lane and Martin (2012).

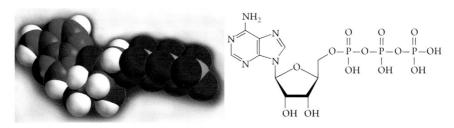

Plate 7.3.   ATP, the energy currency of all life. Illustration adapted from public domain.

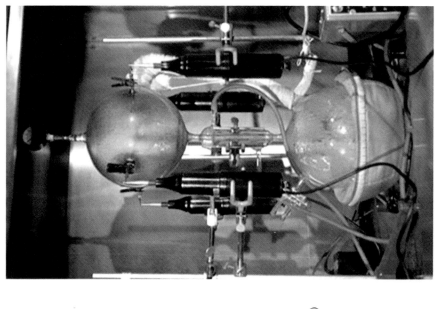

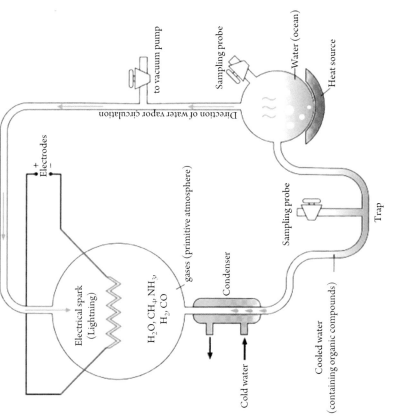

Plate 8.1. Schematic and photo of the Miller apparatus. Illustration adapted from public domain, courtesy of University of California, San Diego.

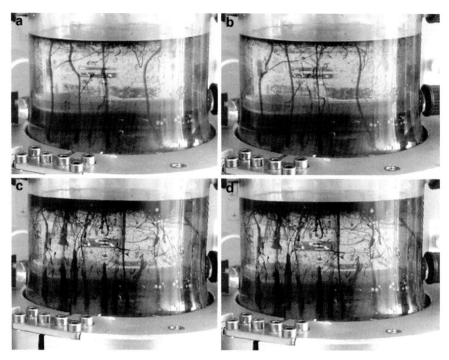

Plate 8.2. Simulation of an alkaline hydrothermal vent (Herschy et al., 2014): a. initial formation of precipitate; b. 20 minutes later; c. one hour later; d. four hours later. Image courtesy of Nick Lane.

EVAPORATION

FLUID AQUEOUS PHASE WITH LIPID VESICLES

REACTANTS ORGANIZED IN 2-DIMESIONAL PLANES OF FLUID LIQUID CRYSTAL

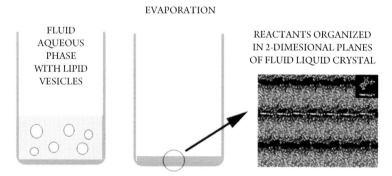

Plate 9.1. Proposed cycle of evaporation and concentration in which an aqueous solvent is replaced by a liquid crystalline solvent. Image on right is a cartoon showing adenosine monophosphate concentrated between the bilayers of a multilamellar phospholipid after evaporation of water and fusion of lipid vesicles. Illustration by author.

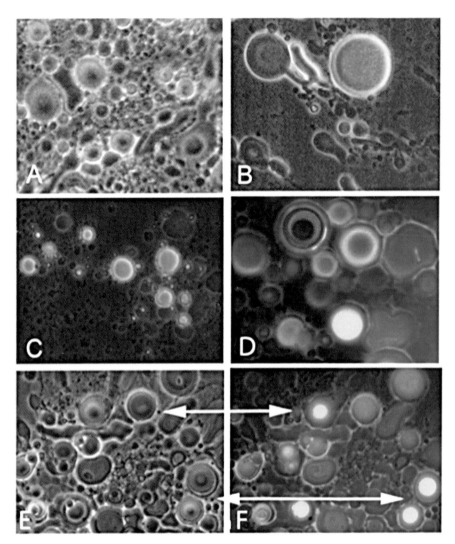

Plate 9.2. Polymeric products are produced, by cycling dAMP + TMP in a lipid mixture, and then accumulate within the membranous vesicles. Panel A shows the mixture before cycling. Panels B, C, and D show fluorescent micrographs of vesicles after 1, 2, and 3 cycles respectively. Panels E and F show phase and fluorescence micrographs of vesicles after 4 cycles. The vesicles were stained with DAPI, a fluorescent dye commonly used to stain DNA in cells. Photographs by author.

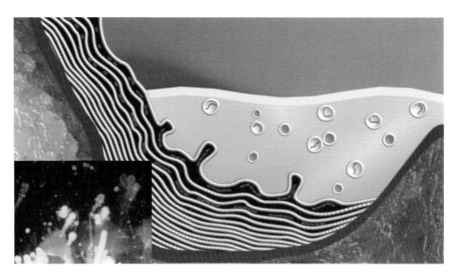

Plate 10.1. Protocell formation and encapsulation of polymers during hydration. Inset: Lipid vesicles with encapsulated DNA (160× original magnification). These were produced by a dehydration–rehydration cycle simulating a small pond in a volcanic hydrothermal site undergoing evaporation and refilling. Protocells containing the DNA are formed when the dried mixture of lipid and DNA is rehydrated. Image adapted from Damer and Deamer, 2015.

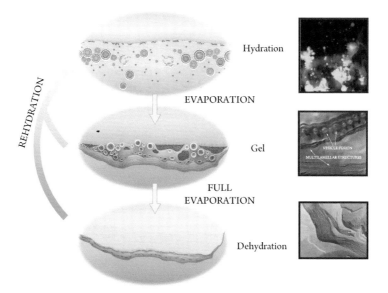

Plate 10.2. An integrated three-phase scenario showing how membranes and polymers can co-evolve in fluctuating hydrothermal conditions on the prebiotic Earth. Image adapted from Damer 2016.

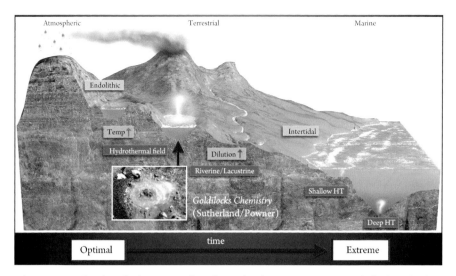

Plate 10.3. A local-scale depiction of a volcanic land mass interacting with fresh and salt water conditions. Credit: Bruce Damer and Ryan Norkus.

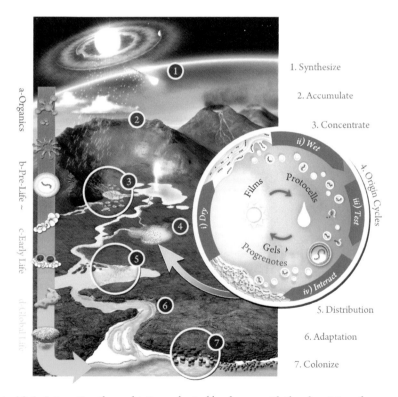

Plate 10.4. Integrating the prebiotic geological landscape with the chemistry of life's origin in hydrothermal fields and subsequent adaptive pathway for early living communities. Credit: Bruce Damer and Ryan Norkus.

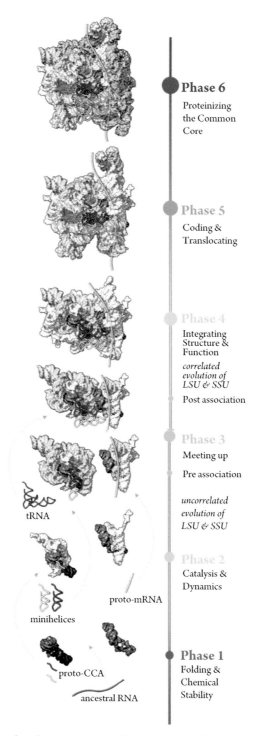

Phase 6
Proteinizing
the Common
Core

Phase 5
Coding &
Translocating

Phase 4
Integrating
Structure &
Function
*correlated
evolution of
LSU & SSU*
Post association

Phase 3
Meeting up
Pre association
*uncorrelated
evolution of
LSU & SSU*

Phase 2
Catalysis &
Dynamics

Phase 1
Folding &
Chemical
Stability

tRNA

proto-mRNA

minihelices

proto-CCA

ancestral RNA

Plate 11.1. Proposed evolutionary path to ribosomes. From Petrov et al. 2015.

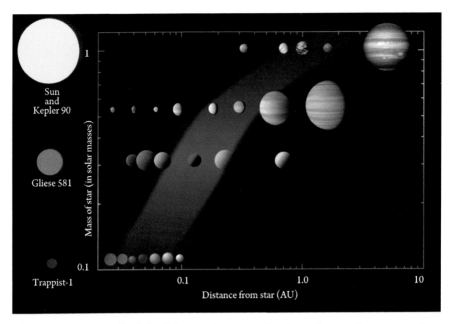

Plate 12.1. Circumstellar habitable zone showing rocky planets in our solar system and Earth-sized planets in orbit around Gliese 581, Trappist-1, and Kepler 90. The entire solar system of Trappist-1 would fit inside the orbit of Mercury. Illustration adapted from public domain images.

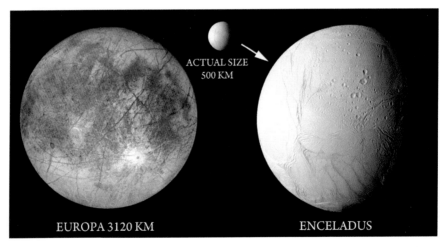

Plate 12.2. Comparison of two icy moons with liquid oceans beneath the ice. Adapted from NASA images, Galileo mission (Europa) and Cassini mission (Enceladus).

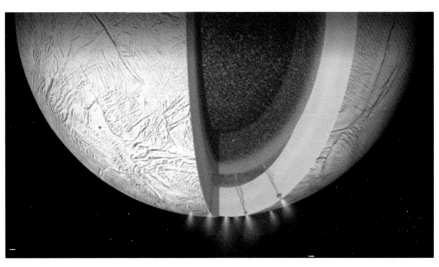

Plate 12.3. Artist's impression of hydrothermal activity in the Enceladus seafloor producing plumes that erupt through the ice into space. Credit: NASA/JPL-Caltech

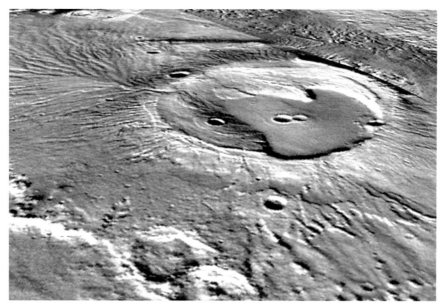

Plate 12.4. Appolinaris Mons (top) and Olympus Mons. The latter is comparable in size to France. Adapted from NASA Mars orbiter missions.

the evolutionary steps leading to life, peptides and RNA formed complexes with novel functional properties beyond those of the individual molecular species. Chapter 11 explores pathways by which this interaction could lead to primitive ribosomes.

The following questions guide the discussion in this chapter:

- What are the roles of condensation and hydrolysis reactions in life today?
- Which compounds have chemical properties that allow them to polymerize?
- What linking bonds could have produced polymers in prebiotic conditions?
- What energy sources were available to drive polymerization in prebiotic conditions?
- How can a steady state be established between polymer synthesis and hydrolysis?
- What functional properties could be exhibited by different polymers?

Condensation Reactions Are the Foundation of Living Systems

A fact that is often not sufficiently emphasized in research on the origin of life is that biopolymers like nucleic acids and proteins are synthesized by condensation reactions. Even the "copying mechanism" that Watson and Crick hinted at in the epigraph is driven by energy made available by condensing a phosphate with an ADP to make ATP. Condensation is defined as a chemical reaction in which the elements of water are removed from between two molecules to form a linking bond. The term "elements of water" is used to make clear that a water molecule is not itself removed. Instead, when the chemical reaction of condensation is written and balanced, one of the products is H_2O, but the actual process involves a more complex mechanism to be described later in this chapter. The opposite of condensation is hydrolysis, in which the elements of water are added to a linking bond in such a way that the bond is broken.

Because of its relative simplicity, it seems reasonable to test condensation as a chemical reaction that could produce polymers required for life to begin. All life today exists in a steady state between polymer synthesis and polymer hydrolysis, with the steady state defined by thermodynamics and kinetics, the two fundamental principles of chemistry. The thermodynamics and kinetics described in this chapter are adapted from Ross and Deamer (2016).

Condensation today is not a direct loss of water to form polymers. Instead, water is subtracted first by the use of metabolic chemical energy or chemiosmotic energy to remove the elements of water from ADP (adenosine diphosphate) and phosphate, thereby forming the high-energy anhydride pyrophosphate bond of ATP. The energy stored in ATP is then used in a variety of ways, one example being phosphate group transfer to activate metabolites:

ATP + glucose → ADP + glucose-6-phosphate

This reaction, catalyzed by hexokinase, chemically activates glucose after it enters a cell so that glucose-6-phosphate can undergo the subsequent reactions of glycolysis.

In another type of reaction, ATP is hydrolyzed to release the free energy in the pyrophosphate bond:

$$ATP + H_2 0 \rightarrow ADP + phosphate + 7\, kcal/mol$$

An important use of the energy released by this reaction is the active transport of ions like Na^+, K^+, H^+, and Ca^{2+} across biological membranes. The transport is carried out by specialized pump proteins embedded in cell membranes. These hydrolyze the ATP and couple the energy to a process in which ions are picked up on one side of the membrane and released on the other. This produces ion concentration gradients that are essential for a variety of cell functions such as muscle contraction and nerve impulse transmission.

However, the synthesis and utilization of ATP as an energy currency was probably a later evolutionary step in the pathway to the first forms of life. The question to be addressed in this chapter is how ester and peptide bond synthesis could give rise to the essential polymers of life in the absence of a highly evolved energy metabolism. Polymers produced by condensation reactions are thermodynamically unstable and would be expected to hydrolyze at a certain rate in aqueous solutions. Therefore, we must find a mechanism by which condensation reactions occur faster than hydrolysis so that polymers will accumulate in a mixture of potential monomers. Although we do not usually think of it in this way, the polymers of all life today exist in a kinetic trap away from equilibrium. It there were no kinetic traps life would not be possible because proteins and nucleic acids would hydrolyze as fast as they are synthesized and could not assemble into living systems.

What Monomers Could Undergo Condensation Reactions and Polymerize in the Prebiotic Environment?

Proteins are composed of amino acids. Because ~70 different amino acids are present in carbonaceous meteorites, an assumption in origins of life research is that dilute amino acid solutions were present on the prebiotic Earth wherever there were bodies of water. Amino acids are defined as any small molecule having both an amine group and a carboxyl group on the same molecule. Even though 70 compounds in meteorites fit this definition, just ten of them are among the 20 amino acids used by life today. The other ten are synthesized by metabolic pathways that appeared later in evolution. From this, we must consider the possibility that primitive life emerged using ten amino acids, which also means that the original genetic code might have required only three nucleobases rather than the four used today. Figure 6.1 shows examples of the four general classes of amino acids based on their physical and chemical properties, including those that have been detected among the organic compounds present in carbonaceous meteorites. The classes include polar, cationic, anionic, and hydrophobic species, which are defined by the properties of the groups attached to the various amino acids. These properties in turn define the biological functions of the proteins synthesized from

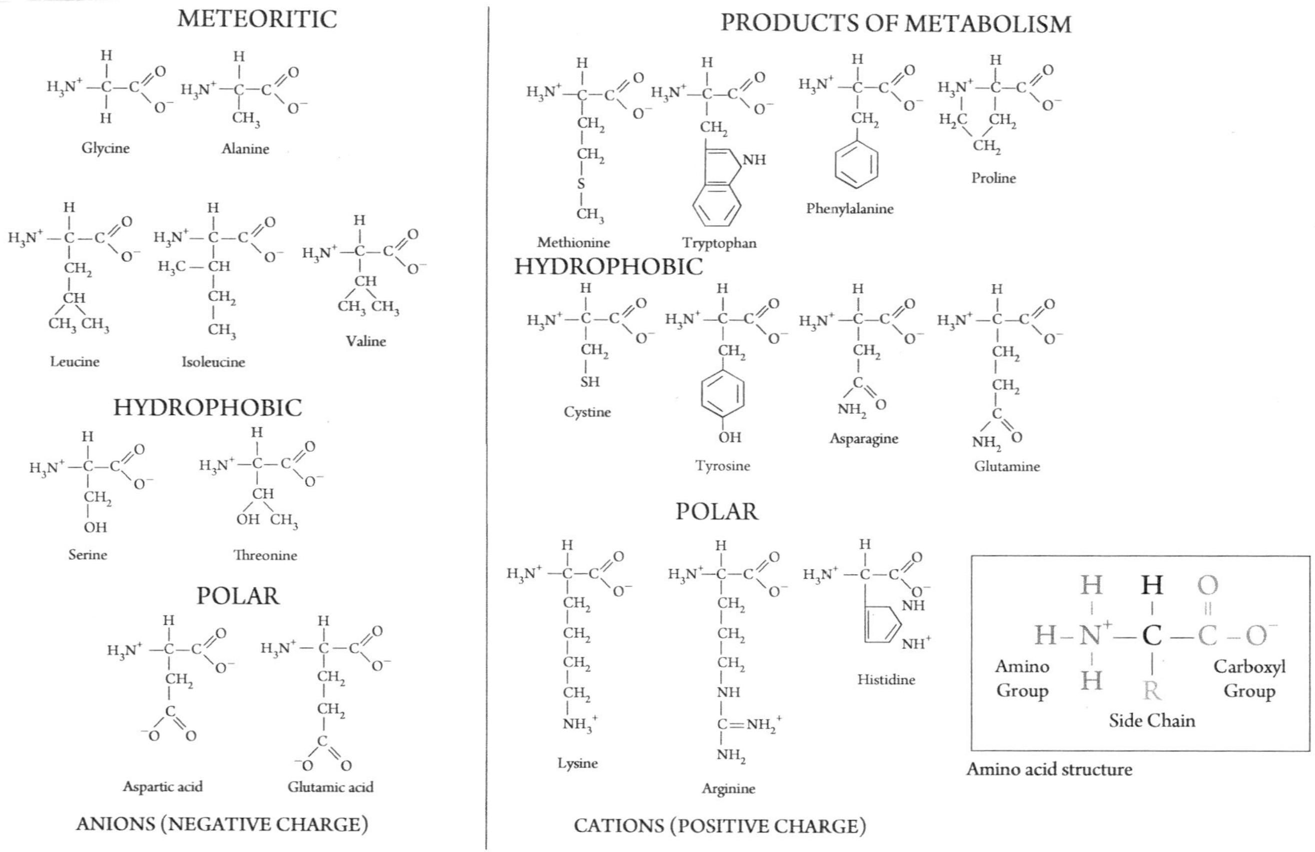

Figure 6.1. The classes and structures of amino acids are shown separated into those found in carbonaceous meteorites on the left and those synthesized in metabolic pathways of life today on the right. The inset shows the basic structure of alpha amino acids, the monomers of proteins.

amino acids. Examples include the catalytic active sites of enzymes and the cytoskeleton that provides structural integrity to cell membranes. Even though such properties are well understood in living cells today, how they emerged in the first forms of cellular life is still unknown.

The monomers of ribonucleic acids are nucleotides composed of the nucleobases adenine, guanine, cytosine, and uracil linked to ribose phosphate through a glycoside bond. In DNA, uracil is replaced by thymine, and ribose is replaced by deoxyribose (Fig. 6.2).

Nucleic acids are considerably more complex than the other biopolymers. Unlike amino acids that can polymerize using peptide bonds, nucleic acids have a sugar-phosphate backbone that is linked by phosphodiester bonds. Simple carbohydrates have been reported to be components of meteoritic organics (Cooper et al., 2001),

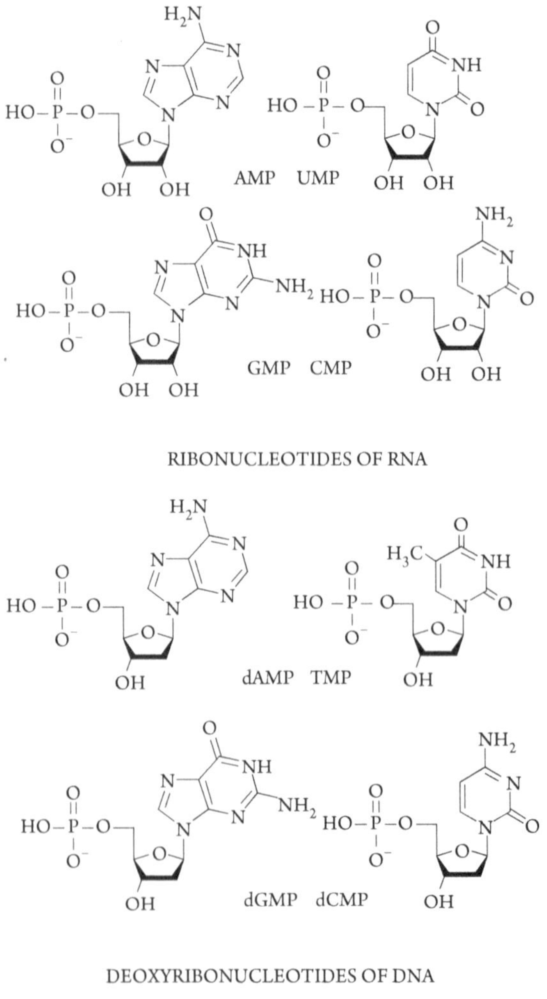

Figure 6.2. The nucleotide monomers of nucleic acids.

and more complex sugars such as ribose are readily produced from formaldehyde by the formose reaction, as described in Chapter 4. Phosphate is among the monomers composing nucleic acids and could have been released in a soluble form from minerals in acidic conditions as described in Chapter 3. The purine and pyrimidine bases are attached to the ribose by an N-glycoside bond, and significant progress has recently been made toward understanding how this bond might be synthesized by a non-enzymatic mechanism. For instance, Nam et al. (2018) showed that N-glycoside bonds form when ribose was mixed in microdroplets with purine and pyrimidine nucleobases then exposed to drying in the vacuum of a mass spectrometer. In a second paper, Kim and Benner (2017) reported that adenine can be directly linked to cyclic ribose phosphate simply by drying. In the rest of this chapter, we will assume that nucleotides were available on the early Earth to participate in the synthesis of nucleic acids.

Esters and Peptide Bonds

Biological ester bonds form between compounds with acidic groups such as carboxylate and phosphate and compounds with hydroxyl (-OH) groups including carbohydrates and glycerol. Chemists have found ways to synthesize esters and peptide bonds in the laboratory, and these reactions are commonly used to produce peptides and nucleic acids with known sequences of amino acids or nucleotides. For example, in peptide synthesis, a chemical group is added to the amino acid monomers that protects against unwanted bond formation. The products are covalently bound to resin beads while specific chemically activated amino acids are added one by one in cycles of reaction and deprotection, and the linking bond is broken when the desired sequence has been synthesized. Needless to say, this process is highly technical with multiple steps of complex reactions and could not occur in nature under prebiotic conditions.

Plate 6.1 illustrates the spontaneous synthesis of an ester bond, the classic reaction of acetic acid with ethanol to produce ethyl acetate, a process called Fischer esterification after Emil Fischer, who with his colleague Arthur Speler first described the reaction in 1895. The reaction is acid catalyzed and usually carried out at temperatures between 60 and 100° C. The second reaction shown is the synthesis of a peptide bond, and the third is the reaction of two mononucleotides to form a dinucleotide linked by a phosphoester bond. All three reactions involve the loss of a water molecule. If we can find conditions that cause water to be drawn away from the reaction, the result will be a continuing synthesis of ester or peptide bonds, in other words, polymerization.

Although the laws of thermodynamics can be used to predict the amount of products in any given reaction when the available free energy and equilibrium constants have been experimentally determined, the rates at which reactions occur are controlled by multiple factors such as temperature, pH, and catalysts, and for the most part can only be established by observation. Here we will deal with the condensation reactions of ester and peptide bond formation, using the ester bond between nucleotides shown in Figure 6.2 as an example.

We can begin by recalling some equations from basic chemistry. A professional chemist will find this explanation elementary, but this is not a chemistry text book. Here, we are trying to make the information and logic clear even to nonchemists.

The condensation reaction between the two nucleotides shown in Figure 6.2 can be written

$$AMP + AMP \iff AMP - AMP + H_2O \qquad (1)$$

where AMP-AMP is the dinucleotide shown in Plate 6.1. In a dilute solution of reactants and products, the equilibrium constant can be measured experimentally and the results plugged into an equation that defines the equilibrium coefficient:

$$Keq = (AMP - AMP)/(AMP) \times (AMP) \qquad (2)$$

where the parentheses indicate concentrations of reactants and products at equilibrium. (A more complete definition will be given later, in which activities are taken into account.)

Because the bulk phase concentration of water does not significantly change during the reaction, it is not included in the definition of Keq. The Gibbs free energy $\Delta G°$ of a condensation reaction is

$$\Delta G° = -2.3RTlogKeq \qquad (3)$$

where R is the gas constant $(8.3\,JK^{-1}mol^{-1})$ and T is the absolute temperature in degrees Kelvin. The units are therefore joules/mol or sometimes cal/mol in which a calorie is 4.1868 joules. The numbers in joules or calories are so large that they are usually expressed as kilojoules (kJ) or kilocalories (kcal).

Now we can perform an experiment in which two reactions are monitored. In the first, we have pure solution of the dinucleotide AMP-AMP and allow it to approach equilibrium by hydrolysis of the ester bond. In the second, we have a pure solution of AMP and allow it to approach equilibrium in the other direction by spontaneous ester bond synthesis as shown in Figure 6.2. When the rates of the two reactions are measured, we find that the hydrolysis reaction proceeds much faster than the synthesis reaction. From the concentrations of AMP and its dimer AMP-AMP at equilibrium we can find Keq by plugging the measured concentrations into equation (2) above, then calculate $\Delta G°$ from equation (3). The free energy of the phosphoester bond calculated from equation (3) is +13.7 kJ/mol (+3.3 kcal/mol). The fact that the energy is positive and expressed in units of heat means that if a 1.0 M solution of AMP-AMP hydrolyzed in one liter of water, the temperature will increase. One kilocalorie will heat a kilogram of water 1° C, so the temperature might be expected to increase by 3.3° C. However, this is an idealized thought experiment; some of the energy would be lost to a change in entropy, so the actual increase in temperature would be less than that.

Now we will write the same equation not for a dilute solution but in a concentrated film of reactants and products produced by evaporation. Reactions are usually performed in solutions by chemists, so at this point we are going beyond ordinary chemical conditions.

$$\{AMP\} + \{AMP\} \iff \{AMP - AMP\} + \{H_2O\} \qquad (4)$$

in which Keq is defined as $\{AMP\text{-}AMP\}\{H_2O\}/\{AMP\}^2$

Instead of concentration, the brackets indicate activities of the reactants and products. Activity is defined as the "effective concentration" of a potential reactant. In the case of water, an example would be the ability of water to hydrolyze an ester bond. The concentration of water in water is 55 M, and in a dilute solution of an ester compound the concentration of water is virtually identical to its activity so that Keq is a constant. But now if $MgCl_2$ is added to the solution to make it 5 M, the water concentration has been reduced because the magnesium chloride uses some of the volume in the mixture. However, the magnesium cation also takes up water in its hydration shell which reduces the effective concentration or activity of water in the mixture and the Keq will be markedly shifted.

In a highly concentrated evaporating film, water becomes one of the reactants and can now be expressed in terms of its activity. Significantly, if a sink for the water is introduced, such as evaporation, the reaction shown in equation (4) is drawn to the right so that dimers and higher oligomers become dominant products. The point of this exercise is that if the activity of water can be reduced in a concentrated film, the free energy of phosphoester bond formation can become exergonic and polymers will be synthesized.

For purposes of illustration, we have used the example of ester-bond synthesis between mononucleotides. However, many studies have also been performed for other linking bonds in biopolymers; the results are shown in Table 6.1, taken from Ross and Deamer (2016)

The standard Gibbs energies of the synthesis of peptide, polynucleotide, and glycosidic links are endoergic, but the energies were all measured in dilute solutions. *Very different results are produced when dilute solutions are evaporated in conditions that produce molecular crowding in an organized state.* Cycles of hydration and dehydration would have been ubiquitous on the early Earth wherever volcanic land masses emerged from the global ocean. Seawater was presumably salty, perhaps even at higher concentrations than today's ocean, but evaporation from the ocean followed by precipitation would supply distilled water to hydrothermal fields on volcanic islands. This process would produce geological conditions analogous to the geysers and hot springs that are abundant in Yellowstone National Park, Iceland, and other areas of active volcanism. An important consideration for the argument presented here is that small pools associated with the hydrothermal fields would undergo continuous cycles of filling and evaporation.

In what follows we will use thermodynamic theory to test whether concentrating the solutes by evaporation is a plausible way to drive prebiotic polymerization reactions.

Table 6.1 **Thermodynamic and Kinetic Parameters Related to Synthesis and Hydrolysis of Biopolymer Linking Bonds**

Thermodynamic and kinetic factors	*Protein, peptide*	*RNA, phosphate ester*	*DNA, phosphate ester*
$\Delta G°$ formation/kcal/mol	2.2	3.3	
Hydrolytic half-lives			
25° C, pH 7	385 yr	5 yr	0.2 Myr
85° C, pH 2.5	7 min	8 days	46 yr

This idea is obvious and was proposed years ago as a way to synthesize prebiotic macromolecules (Lahav et al., 1978). Table 6.1 also shows the hydrolytic half-life of each linkage and there is an important point to be made concerning the effect of temperature and pH on this. At ordinary temperatures and neutral pH, the bonds are surprisingly stable, with half-lives measured in years. However, at the temperature and pH ranges assumed to prevail on the prebiotic Earth, peptide bonds are hydrolyzed in minutes and RNA ester linkages in days.

Concentration and Water Activity Can Drive Condensation Reactions

How can concentration of reactants by evaporation cause ester and peptide bonds to be synthesized? It is useful to recall the fundamental equation that relates the free energy available in any given chemical reaction:

$$\Delta G = \Delta H - T\Delta S \qquad (5)$$

where ΔG is the change in Gibbs free energy, ΔH is the change in enthalpy and ΔS is the change in entropy. At equilibrium, ΔG is zero and the forward and reverse reaction rates are equal. In any condition where reactants exist in a state away from equilibrium, ΔG by convention is negative, meaning that energy is available and then released as the reaction proceeds back to equilibrium. It follows that in order to push a reaction away from equilibrium, energy must be added to the reactants. There are two ways to do this. We can find conditions in which the water activity shown in equation (4) is reduced, so that the reaction is pulled to the right toward polymerization. Or, an external energy source can produce decreased entropy in the system, which would move ΔG toward a negative value, for instance, by introducing increased order into the potential reactants.

We will now return to the example illustrated in equation (1), the equilibrium of ester bond formation between two ordinary mononucleotides:

$$AMP + AMP \Leftrightarrow AMP - AMP + H_2O \qquad (6)$$

Is it possible that simply concentrating the reactants by evaporation can drive the system toward extensive polymerization? The answer is no. If we assume +3.3 kcal/mol for the formation of phosphate esters at 85° C, it would require an impossible concentration of 300 M just to shift the equilibrium to the point where there are more dimers than monomers.

If concentration alone is insufficient, other factors associated with the concentrating effects must be considered. One such factor is molecular crowding (Kim and Yethiraj, 2009) which affects the activity of water. When the solute concentration increases and approaches an anhydrous state and molecular crowding approaches a maximum, it is obvious that cavities will be left behind as water leaves. The average size of the cavities is related to water activity, as shown in Figure 6.3, and water activity can be used to calculate the change in Gibbs free energy. The surprising result is that the free energy

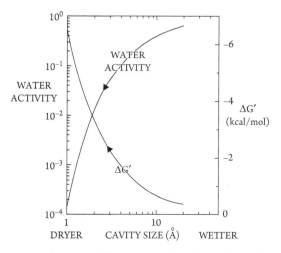

Figure 6.3. The activity of water and the corresponding favorable Gibbs energy change at 85° C as a function of the characteristic cavity size (Ross and Deamer, 2016).

available to drive condensation reactions changes from near zero to over -5 kcal/mol. *In other words, the energy put into the system to evaporate an aqueous solution of potential reactants to near dryness is conserved as a favorable change in free energy available to drive condensation reactions.*

Favorable Changes in the Entropy Term

Decreasing water activity by evaporation is one contribution to the free energy available to drive condensation reactions, but changes in entropy could also contribute. Keep in mind that the entropy of a dilute solution of nucleotides is at a maximum because the solutes are randomly dispersed in the solution. Now consider what might happen to the entropy of the system as evaporation increasingly concentrates the solutes. In the case of a solute that can form crystals, the result is obvious. As the solution dries the solutes pass through a concentration of maximum solubility and then begin to crystallize. When completely dry, the crystalline solutes have by definition reached a maximum degree of ordering and minimal entropy.

    However, solutions of pure nucleotides in prebiotic conditions are implausible, so we need to consider what would happen with solutions of mixed nucleotides. If the nucleotides just formed a disordered glass during evaporation, the entropy of the system would decrease by a small amount related to a large volume of dilute solution being reduced to a small volume of concentrated solution. But what if self-assembly of the nucleotides brings additional order to the system? In fact, in the presence of a lipid matrix Toppozini et al. (2013) observed that AMP becomes ordered between lamellae of the multilamellar lipid matrix and formed paracrystalline arrays. Himbert et al. (2016) extended this to mixtures of AMP and UMP undergoing dehydration and reported that the nucleotides formed orderly stacks in a variety of conditions.

There has been no attempt yet to determine quantitatively how much entropy is changed by these ordering processes, but it is clear that the change will be favorable as order is imposed by multilamellar structures and self-assembly associated with base stacking.

Theoretical Foundations of Polymerization Driven by Evaporation

Given this introduction, we can now test the thermodynamic and kinetic basis of polymer synthesis driven by evaporation and concomitant concentration of reactive solutes (Ross and Deamer, 2016). The test was performed with Kintecus software, using a hydrolysis rate constant of 10^{-6} s^{-1} at pH = 2.5 determined experimentally for the hydrolysis of uridylyl (3'-5') uridine at 90° C (Oivanen et al., 1998). We assume that a certain amount of free energy was made available by a decrease in water activity and a favorable entropy change, as just described. If we take the ΔG value as a negative 10 kcal/mol the system of monomers rapidly shifts to a mixture of oligomers in which *elongation is energetically favored* and each oligomer in the concentrated mix is more stable than its immediate precursors (Fig. 6.4). Moreover, the time to equilibrium has fallen to a few tens of seconds, so the uncatalyzed kinetics of the reaction can be surprisingly fast, at least in this idealized computational simulation.

The reaction modeled in Figure 6.4 reflects just a single dehydration cycle and ignores a significant physical fact that would otherwise limit oligomerization. When

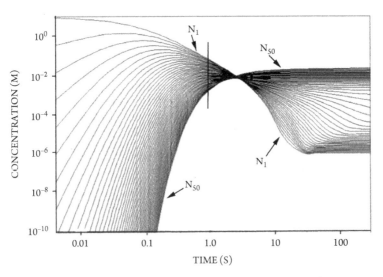

Figure 6.4. Numerical simulations of the evaporation and rehydration of 6.6 mM solutions of mononucleotide concentrated by a factor of 1400 and ΔG = −10 kcal/mol. The system comes to equilibrium in approximately 30 seconds, and oligomers dominate the products. (Ross and Deamer, 2016)

a solution of mononucleotides dries, the solutes would be expected to go from a solution allowing free diffusion of reactants to a solid state in which diffusion ceases. In the absence of diffusion, elongation of polymers is inhibited, and the predominant product would be a small yield of dimers that form between neighboring monomers in the dry state.

This brings us to the major point that is the foundation of the concepts being advanced in this book: *if a lipid is present, the monomers do not get trapped in a solid state. Instead they are concentrated and organized within a multilamellar matrix of a liquid crystal in which they are able to diffuse within a two-dimensional plane.* In other words, evaporation simply exchanges a polar fluid solvent—water—for the less polar fluid solvent of a liquid crystal. Furthermore, the polymerization reaction does not occur just once and reach equilibrium but instead proceeds in multiple cycles of hydration and dehydration.

Kinetic Traps

The polymers of life are assembled from amino acids and nucleotide monomers that are highly soluble in water, and their polymerization starts with the monomer-to-dimer conversion and continues on to the production of very large oligomers:

$$nA \; \rightleftarrows \rightleftarrows \rightleftarrows \; A_n + (n-1)H_2O \tag{7}$$

For the process to proceed successfully, however, it is essential for the reactions in the growth direction to be energetically downhill, leading to product ratios, A_{n+1}/A_n greater than 1. A ratio less than 1 consigns the process to no more than the first few short oligomers, and life cannot develop.

For the case of commercial polymer manufacture—polyethylene and polystyrene production for example—ratios greater than 1 readily arise since the reactions are very highly energetically downhill. For life's polymers, the opposite is true. The equilibria in equation (6) lie heavily to the left, so two questions must be answered before we can understand possible polymerization reactions in prebiotic conditions. What conditions can shift the reactions to an energetically downhill direction and move the equilibria to the right, and how can the polymers be sustained in the face of the inherent thermodynamic drive favoring hydrolysis in aqueous solutions?

There are two answers to these questions, and the first is easy to understand. If chemical free energy is used to drive a condensation reaction in the dehydrated phase of a wet–dry cycle, this reaction can happen very quickly in the highly concentrated film as illustrated in Figure 6.4. However, when the film is relaxed by dissolving it in the wet phase of a cycle, the rate of hydrolysis can be very slow, so it is inevitable that polymers will accumulate.

The second answer is related to changes in the nature of water and the structures of large organic solutes when solutions become very concentrated. The changes are represented symbolically in Figure 6.5 which charts the energies of the reaction components along the reaction path. Because water is an excellent solvent for monomers like amino acids and nucleotides, the high level of stability provided by the

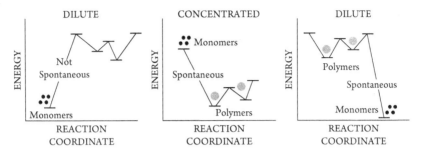

Figure 6.5. The basis of a kinetic trap. A dilute solution of potential monomers cannot form polymers because the reaction is thermodynamically uphill. However, during drying, the monomers are concentrated and free energy becomes available to drive rapid polymerization. When the dry polymer products are dissolved again in water, forming a dilute solution, they are trapped in wells that have an activation energy barrier to hydrolysis, so the downhill back reaction is much slower.

solvent and solute maintains a product ratio $A_{n+1}/A_n \ll 1$; polymerization is accordingly nonspontaneous and does not occur in dilute solutions.

At high concentrations the energetics are inverted and oligomer formation shifts to an energetically downhill condition. The swing comes about because the free volume of water falls to very low values and begins to approximate the volume of the solutes themselves. As a consequence, its solvent qualities decline and the solutes move toward lower energy states that shift equation (6) to the right, easing the energetic liability both through the production of more water and the reduction of solute volumes. This readjustment in response to changing conditions is known as Le Chatelier's principle. Finally upon dilution, hydrolysis of the polymers is also spontaneous but the rate is much slower than the rate of the condensation reaction.

On this basis, one proposal for the origin of life involves evaporating ponds as sites for the very first formation of biopolymers. On the path to life, however, the polymers must at some point return to relatively dilute aqueous solutions, and the challenge of the second question then emerges. It would appear at the outset that the polymers would simply revert back to their monomers as rapidly as they had been formed, and in the end, evaporation would have been a futile exercise. The reversal is only partial, however, and therein lies the resolution to the issue as is shown in the second and third portions of Figure 6.5. Some fraction of the product polypeptide or polynucleotide chains formed in the concentrated state would have been large enough to fold into compact secondary structures, shielding the hydrolytically sensitive peptide and ester groups within the interiors of the structures from reaction with water. The folded structures provide a molecular geometry allowing additional stability through energy-lowering internal hydrogen bonds and van der Waals interactions.

Depending on their sizes and shapes, the folded polymers would fall into a series of stability wells of varying depths. The wells are referred to as kinetic traps: relatively low-energy havens allowing survival of the more stable fraction of the oligomer population. Relatively unstable oligomers return to monomers, and what then emerges during dilution/concentration cycles is a dynamic filtering out of the unstable structures and an enrichment of stable structures. For nucleotides, for example, the addition of a monomer to lengthen a chain results in a ~3 kcal/mol penalty. However, each hydrogen-bonded pair that can form between complementary bases in the oligomer leads to a 1–2 kcal/mol benefit, and two or more pairings within the chain will then begin to provide some level of permanency. Subsequent pairings still farther along the chain afford additional stability and this penalty–benefit tradeoff ultimately results in a more-or-less Darwinian-like selection, on a molecular level, of stable structures that survive chemical and physical stresses.

Specifically, the well depth represents an activation energy, Ea, and slows hydrolysis in accord with the kinetic equation $k = Ae^{-Ea/RT}$. The deeper the well, the greater the activation energy barrier and the slower the hydrolysis. We emphasize again that condensation must be thermodynamically downhill for life to have arisen and to continue. The reality of kinetic throttling of hydrolysis permitted protocell populations to develop catalysts that selectively and cooperatively promoted condensation reactions by lowering activation energies. Over time, microbial populations could then emerge that used metabolic energy to drive polymerization reactions.

The Effect of Cycling

Most reactions studied by chemists and biochemists have a starting mixture of activated reactants with free energy available so that the reaction spontaneously runs to an equilibrium with defined products. An important exception to this rule is the polymerase chain reaction (PCR) in which a small number of specific DNA molecules are amplified by a million-fold or more. In this case, the activated reactants are four nucleoside triphosphates, and the replication reaction is catalyzed by the *taq* polymerase isolated from a thermophilic bacterium, *Thermus aquaticus*, from which the enzyme gets its name. If the reaction is run just once, it reaches completion when all the single-stranded DNA template molecules have been replicated to form duplex strands. The ingenious idea first put forward by Kary Mullis is to run the reaction multiple times using cycles of heating and cooling. Each heat cycle melts the duplex strands so that single-stranded products can be replicated again when the mixture is cooled and the polymerase goes to work. If twenty cycles are used, the original DNA is amplified more than a million-fold with the result that a few picograms of template become a few micrograms of product. *The wet–dry cycles in the pools of a hydrothermal field can be viewed as a natural version of PCR.* However, products are not being replicated by a catalyzed synthesis using activated reactants, but instead accumulate as polymers in the kinetic trap described earlier.

We will spend the rest of this chapter discussing what happens in the hydration phase of a wet–dry cycle and the influence of cycling on the accumulation of oligonucleotides. For purposes of illustration, imagine a condensation reaction in which an ester bond is synthesized at a certain rate during evaporation but then undergoes hydrolysis at the same rate when water is added back to the mixture. Under these conditions it is obvious that extensive polymerization cannot occur. And yet, experimental measurements demonstrate that oligomers do accumulate during multiple wet–dry cycles, which means that the rate of the condensation reaction must exceed the rate of hydrolysis.

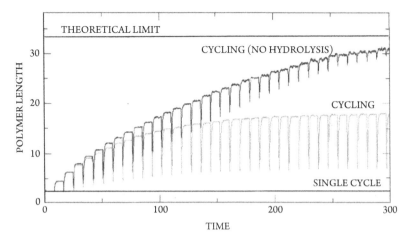

Figure 6.6. Mean chain length of oligomers produced during cycling of a theoretical reaction mixture. Time is given in arbitrary units. Adapted from Higgs (2016).

Is this possible? Of course it is. As noted earlier in this chapter, all life depends on the fact that polymers are synthesized by condensation reactions faster than they are hydrolyzed, and this is referred to as a kinetic trap. The same is true of the reaction shown in Figure 6.5, in which oligomers are synthesized in seconds, while the hydrolytic half-life of the phosphoester bond shown in Table 6.1 at pH 2.5 and 85° C is eight days. Despite the existence of a kinetic trap, in practice condensation reactions cannot produce indefinitely long polymers: as chain length increases, the probability also increases that a single hydrolysis reaction will break one of the many ester linkages in a longer molecule with the result that shorter molecules are produced.

Paul Higgs (2016) has performed numerical modeling of the effect of cycling on a polymerization reaction occurring in confined spaces. The calculation incorporated the number of cycles, with or without diffusion or hydrolysis, and the result was presented as mean length of polymers synthesized. One set of results shown in Figure 6.6, make clear that a single cycle has a very limited yield of dimers, while polymers of increasing length accumulate with cycling. This numerical simulation is consistent with experimental results described in Chapter 9.

Peptide Bonds

Condensation of amino acids into oligopeptides has been demonstrated by Bernd Rode (1999) and more recently by Lee Cronin and coworkers (Rodriguez-Garcia et al. 2015). For instance, Rode found that 3 M sodium chloride effectively acts as a dehydrating agent to overcome the thermodynamic barrier of peptide bond formation in evaporating aqueous solutions, perhaps because the first hydration shell of the sodium ion was assumed to no longer be saturated with water molecules (Jakschitz and Rode, 2012).

Although these results show that peptide bond formation can be driven by simple cycles of dehydration and rehydration, at some point polymerization reactions must have involved a more complex mechanism of chemical activation occurring in aqueous solutions. For many years, research has focused on discovering a plausible condensing agent that can perform this feat, and carbonyl sulfide (COS) is a likely candidate. COS is a reactive compound that has been detected in volcanic gas and mineral ash along with its chemical relatives, carbon disulfide and carbon dioxide. Leman et al. (2004) found that if COS is present in an aqueous solution of amino acids, di- and tripeptides are synthesized with yields up to 80%. In a second paper Leman et al. (2006) reported that amino-acyl phosphate anhydrides were synthesized, with yields of up to 30%, in mixtures of amino acids, phosphate, and COS.

These results offer a clue to the manner in which phosphate was initially incorporated into primitive metabolic pathways, particularly those leading to peptide bond formation and the synthesis of small oligopeptides that may have served as catalysts and structural components of early life.

Conclusions and Open Questions

To summarize, liquid crystal microenvironments are able to organize monomers within a lipid matrix when phospholipid vesicles are mixed with mononucleotides and dried. If the monomers are nucleotides, long strands of RNA-like molecules are synthesized by a condensation reaction when the reactants are exposed to one or more cycles of dehydration and elevated temperatures, followed by rehydration. The chemical potential driving the reaction is supplied by the anhydrous conditions in which water becomes a leaving group, with heat providing activation energy.

Polymerization of oligomeric nucleic acids within liquid crystalline matrices is a novel approach, so a number of questions must be addressed in order to better understand the reaction. The experimental studies to be described in Chapter 9 have been limited to a narrow range of conditions that happen to work: phospholipid matrices as organizing agents, acidic pH ranges, carbon dioxide atmosphere, wet-dry cycles, elevated temperature ranges, and low ionic strength. These are all experimental variables, and future research will explore a much broader range. The conditions described here simulate the kinds of wet–dry cycles at elevated temperatures that would have commonly occurred in geothermal sites on the early Earth. Furthermore, there is a growing consensus that self-assembled lipid membranes can provide the compartments necessary to maintain systems of polymeric catalysts in the evolutionary pathway leading to the origin of cellular life. The fact that an organizing matrix of lipid multilayers can promote polymerization suggests a robust mechanism by which relatively long strands of oligomeric nucleic acids could have been synthesized and then encapsulated in microscopic protocells. Each cellular compartment would contain a different set of polymeric products, representing an experiment in a natural version of combinatorial chemistry.

Degradation and Decomposition

Although synthesis of ester bonds linking mononucleotides into oligomers has been demonstrated experimentally, competing decomposition reactions must be taken into account. In the case of nucleotides and their polymers, the reactions include hydrolysis of ester bonds, depurination of purine nucleotides, and deamination of cytosine to uracil. Mungi and Rajamani (2015), for example, observed significant levels of depurination in hydrothermal simulations at ranges below pH 3. Carbohydrates are subject to chemical damage through the Maillard reaction and caramelization, producing the familiar brown polymers of baked and grilled food. The question is whether chemical decomposition of monomers occurs at rates that would inhibit polymer synthesis in hydrothermal conditions. Because living systems did in fact assemble in prebiotic conditions, the answer is yes, but further research is required before definitive answers can be given.

An important perspective is that the reactions described here do not happen just once in a system approaching equilibrium, but multiple times in cycles. Even though degradation does occur at a certain rate, this is balanced by continuing cycles of synthesis. In the specific case of nucleic acids, the fact that monomers and polymers are sufficiently stable to withstand hydrothermal conditions is made abundantly clear by the PCR reaction in which reactants and products in an aqueous solution are exposed multiple times to the temperature ranges characteristic of hydrothermal fields. There must certainly be degradation of the reactants and products, but these are negligible even after 40 or more cycles of 90-degree temperatures.

Relation to Hydrothermal Vent Scenario

The entire thrust of the argument presented in this chapter is that evaporation, molecular crowding, and reduction in water activity are essential thermodynamic drivers for prebiotic polymer synthesis. If this is correct, polymer synthesis seems unlikely to be possible in hydrothermal vents in which such conditions are absent.

References

Cooper C, Kimmich N, Belisle W, Sarinana J, Brabham K, Garrel L (2001) Carbonaceous meteorites as a source of sugar-related organic compounds for the early Earth. *Nature* 414, 879–883.

Fischer E, Speier A (1895). Darstellung der Ester. *Chemische Berichte* 28, 3252–3258.

Higgs PG (2016) The effect of limited diffusion and wet–dry cycling on reversible polymerization reactions: Implications for prebiotic synthesis of nucleic acids. *Life* 6, 24–30.

Himbert S, Chapman M, Deamer DW, Rheinstadter MC (2016) Organization of nucleotides in different environments and the formation of pre-polymers. *Sci Rep* 6: 31285.

Jakschitz TA, Rode BM (2012) Chemical evolution from simple inorganic compounds to chiral peptides. *Chem Soc Rev* 41, 5484–5489.

Kim JS, Yethiraj A (2009) Effect of macromolecular crowding on reaction rates: A computational and theoretical study. *Biophys J* 96, 1333–1340.

Kim HJ, Benner SA (2017) Prebiotic stereoselective synthesis of purine and noncanonical pyrimidine nucleotide from nucleobases and phosphorylated carbohydrates. *Proc Natl Acad Sci USA* 114, 11315–11320.

Lahav N, White D, Chang S (1978) Peptide formation in the prebiotic era: Thermal condensation of glycine in fluctuating clay environments. *Science* 201, 67–69.

Leman L, Orgel L, Ghadiri MR (2004) Carbonyl sulfide-mediated prebiotic formation of peptides. *Science* 306, 283–286.

Leman LJ, Orgel LE, Ghadiri MR (2006) Amino acid dependent formation of phosphate anhydrides in water mediated by carbonyl sulfide. *J. Am. Chem. Soc.* 128, 20–21.

Mungi CV, Rajamani S (2015) Characterization of RNA-like oligomers from lipid-assisted nonenzymatic synthesis: Implications for origin of informational molecules on early Earth. *Life* 5, 65–84.

Nam I, Nam HG, Zare RN (2018) Abiotic synthesis of purine and pyrimidine ribonucleoside in aqueous microdroplets. *Proc Natl Acad Sci USA* 115, 36–40.

Oivanen M, Kuusela S, Lonnberg H (1998) Kinetics and mechanisms for the cleavage and isomerization of the phosphodiester bonds of RNA by bronsted acids and bases. *Chem Rev* 98, 961–990.

Rode BM (1999) Peptides and the origin of life. *Peptides* 20, 773–786.

Rodriguez-Garcia M, Surman AJ, Cooper GJT, Suárez-Marina I, Hosni Z, Lee MP, Cronin L (2015) Formation of oligopeptides in high yield under simple programmable conditions. *Nature Comm* 6, 8385.

Ross DS, Deamer D (2016) Dry/wet cycling and the thermodynamics and kinetics of prebiotic polymer synthesis. *Life (Basel)* 6, 28

Toppozini L, Dies H, Deamer DW, Rheinstädter MC (2013) Adenosine monophosphate forms ordered arrays in multilamellar lipid matrices: Insights into assembly of nucleic acid for primitive life. *PLoS ONE* 8, e62810.

Bioenergetics and Primitive Metabolic Pathways

With the metabolic pathway alone, you have a very good starting point for life, but it is not life, just a chemical-reaction network. You also need things like membranes to contain the reactions, and the genetic machinery that enables inheritance. How do you bring these elements together in one environment and in non-extreme conditions, and make them work? This is still a big challenge.

Markus Ralser, 2017

Overview and Questions to be Addressed

It seems inescapable that at some point primitive cells incorporated chemical reactions related to what we now call metabolism. In all life today, metabolic reactions are driven by sources of chemical or photochemical energy, and each step is catalyzed by enzymes and regulated by feedback systems. There have been multiple proposals for the kinds of reactions that could have been incorporated into early life, but so far there is little consensus about a plausible way for metabolism to begin. This chapter will briefly review the main ideas that are familiar to chemists as solution chemistry but then ask a new question from the epigraph: how can reactions in bulk aqueous solutions be captured in membranous compartments? This question is still virtually unexplored, but I can offer some ideas in the hope of guiding potentially fruitful approaches. Because metabolism is such a complex process, it is helpful to use bullet points to help clarify the discussion. The first is a list of questions that guide the discussion, the second is list of facts to keep in mind, and the third is a list of assumptions that introduce the argument.

Questions to be addressed:

- What are the primary metabolic reactions used by life today?
- What reactions can occur in prebiotic conditions that are related to metabolism?
- How can potential nutrient solutes cross membranes in order to support metabolism?
- How could metabolic systems become incorporated into primitive cellular life?

Primary Metabolic Reactions in Life Today

Metabolism can be defined as the activity of catalyzed networks of intracellular chemical reactions that alter nutrient compounds available in the environment into a variety of compounds that are used by living systems. Most of the reactions are energetically downhill, so there is an intimate association between the energy sources available to life and the kinds of reactions that can occur. Here is a summary of energy sources used by life today:

- Light is by far the most abundant energy source, totaling 1360 watts per square meter as infrared and visible wavelengths.
- Chemical energy in the form of reduced carbon compounds is made available by photosynthesis.
- Oxidation–reduction potentials between reduced and oxidized compounds provide energy for electron transport reactions embedded in membranes.
- Chemiosmotic energy is developed by proton transport coupled to electron transport in microbial membranes, mitochondria, and chloroplasts.

We can begin by considering metabolic reactions used by all living cells today, then consider which chemical reactions in the prebiotic environment might be incorporated by early forms of life. Biochemists are familiar with the chemistry of metabolism, but many readers of this book will not have studied biochemistry, so it is worth outlining the primary metabolic pathways.

Most of the energy driving metabolism in the biosphere is supplied by sunlight coupled to photosynthesis. Light energy is harvested by pigment systems, usually involving chlorophyll, which capture the energy in order to synthesize ATP and produce the reduced form of NADPH (Plate 7.1). The chemical energy available in these two molecules is in the anhydride pyrophosphate bond between the second and third phosphate groups of ATP and in the reducing power of NADPH (Fig. 7.1).

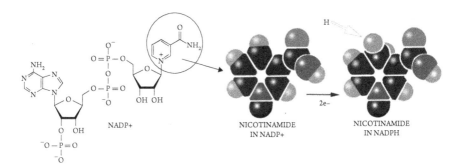

Figure 7.1. The reducing power in NADPH comes from a covalent bond composed of two electrons—attaching a hydrogen atom to the reduced NADPH, indicated by the open arrow—that can be donated to other molecules. NADPH has the same structure as NADH in mitochondria except for one extra phosphate added to the ribose of the ADP molecule.

Nicotinic acid is synthesized by plants and is then attached to ADP to form nicotinamide adenine diphosphate (NADP+). When NADP+ is reduced by the light reactions
of photosynthesis to form NADPH, an extra hydrogen atom is linked to the ring along
with two electrons that form the covalent bond (indicated by the open arrow in Fig. 7.1).

It is a remarkable fact that virtually all of the energy driving life in the biosphere
begins with those two electrons. The source of the electrons is the water-splitting reaction of photosynthesis, and the electrons are ultimately used to reduce CO_2 to form
biochemicals such as glucose, amino acids, and fatty acids (Plate 7.1). Significantly, because water is the ultimate source of the electrons used in photosynthesis, molecular
oxygen is released as a byproduct of photosynthesis.

The microbial invention of a photochemical process that synthesized reduced biochemical compounds from CO_2 and molecular oxygen from water led to a second
major source of energy beginning about 2 billion years ago, which involves the transport of high energy electrons in reduced compounds back to oxygen. This is analogous
to the chemical energy in a flashlight cell that produces a voltage and usable electrical
current. The energy of sunlight charges the cell and the stored energy is used to generate a proton gradient across the membranes of microorganisms, mitochondria, and
chloroplasts, a process called chemiosmosis.

The third source of metabolic energy is chemical energy stored in more complex
molecules during their synthesis, which can be released when the larger molecules are
broken down into smaller molecules. Today's life makes use of this thermodynamic fact
and evolved enzymatic catalysts that released energy stored in reduced compounds, such
as breaking down 6-carbon glucose into smaller, 3-carbon molecules, a process called
glycolysis. Phosphate from 2 ATP is required to activate glucose and intermediates in
the first steps of glycolysis, but 4 ATP are synthesized at two points in the second phase,
so glycolysis alone is a source of energy in the form of ATP:

Glucose (six carbons) + 2 ATP → 2 three carbon intermediates
→ 2 pyruvic acid + 4 ATP + 2 NADH

Some of the pyruvic acid is used in a series of later reactions to synthesize amino acids
and fatty acids, while the rest proceeds into a remarkable oxidative process called the
citric acid cycle in which electrons are stripped from the intermediates and ultimately
delivered to electron transport enzymes. As these enzymes transport the electrons to
molecular oxygen, they have the unique ability to drive proton transport across the
membranes in which they are embedded. Because the protons are positively charged,
the result is a membrane potential of approximately 200 millivolts.

The last step of energy coupling is extraordinary. The protons that are pumped
across the membrane produce a membrane potential but can move back through the
membrane if there is a pathway available. The pathway is provided by a complex enzyme called ATP synthase which has a channel through which protons can flow. The
flow is not simply a leak, but instead drives the rotation of a head group that spins 130
times per second. The head group has binding sites for ADP and phosphate, which
when they enter the sites, combine to form ATP. However, the ATP remains bound
tightly to the head group and this is where the rotation is essential because it activates

a mechanism that releases the ATP and resets the enzyme so that it can bind another ADP and phosphate. The net result is the synthesis of 34 ATP molecules for every pair of electrons that travels from pyruvic acid to oxygen.

The biochemical reactions of metabolism are learned by undergraduate students majoring in biochemistry. We tend to accept them as common knowledge, forgetting that a century ago they were completely mysterious. It required the focused efforts of hundreds of talented biochemists to establish the networks of metabolic processes. As the details of the intricate pathways slowly emerged, we began to realize how complex the machinery of life actually was, and well-deserved Nobel prizes were awarded to Hans Krebs for elucidating the citric acid cycle, to Melvin Calvin for discovering the dark reactions of photosynthesis, to Peter Mitchell for positing the chemiosmotic theory, and to Paul Boyer for explaining the rotation of ATP synthase. Today we take this knowledge for granted, but now we face another enormous gap of ignorance: How could such complexity begin 4 billion years ago when the first primitive life originated?

Prebiotic Chemistry is the Forerunner of Today's Metabolism

In what follows, I will briefly outline the main ideas that have been proposed by other researchers over the years to bridge this gap, but with an important caveat: None of these ideas has advanced very far beyond the conjecture stage. There have been a few attempts to test them in the laboratory, but in my judgment, the evidence has not yet led to a consensus explaining how prebiotic chemical reactions could be incorporated into primitive metabolism. Here is a list of potential energy sources on the prebiotic Earth:

- Light energy would be the most abundant energy source.
- Chemical energy would be present in localized volcanic sites.
- Redox reactions can occur as electrons flow from reduced to more oxidized compounds.
- Heat provides activation energy, as well as concentrating potential reactants by evaporating dilute solutions.

Geochemical Reactions

The environment of the prebiotic Earth was far from equilibrium, so multiple chemical reactions were occurring simultaneously. How could certain specific reactions be sorted out and used by the earliest forms of life? Harold Morowitz, served as a faculty member at George Mason University until his death in 2015, and in his book *Beginnings of Cellular Life* (1992) he considered how prebiotic chemistry could evolve into the metabolic pathways of today's life. Morowitz presented two novel ideas, that the origin of life required cellular compartments, and that ancient metabolic pathways can be deduced from patterns in biochemical compounds and metabolic reactions used by contemporary life. Morowitz put bioenergetics front and center, beginning with the capture of sunlight:

> The principal route of energy flow in contemporary biology is from solar flux to: (1) an oxidized and reduced AH_2 and O_2, and (2) the formation of a pyrophosphate or polyphosphate. The chemical energy that feeds into metabolic processes to drive the rest of biochemical action consists of reducing power and phosphate bond energy. This statement is one of the most powerful generalizations of modern biochemistry, and there is every reason to believe it was primitive.

Robert Hazen and George Cody work at the Carnegie Institution of Washington, just across the Potomac River from George Mason University, and they were inspired by Morowitz's insights to undertake new experiments (Morowitz et al., 2000; Cody et al., 2006). Their approach focused on the possibility that the citric acid cycle, now central to metabolic processes in cellular life, could be driven in reverse under hydrothermal conditions in a series of reactions catalyzed by nickel and iron sulfide. For example, they demonstrated that in the presence of CO, an unsaturated hydrocarbon (1-nonene) can undergo a series of reactions that ends up with hydroaconitic acid, a tricarboxylic acid similar to citric acid. The authors concluded that "These results point to a simple hydrothermal redox pathway for citric acid synthesis that may have provided a geochemical ignition point for the reductive citrate cycle." Cooper et al. (2011) observed that some of the compounds required for the citric acid cycle were in fact present in the mixture of organics in the Murchison meteorite.

The most recent attempt to determine whether the reactions of the citric acid cycle could occur in the prebiotic conditions was undertaken by Keller et al. (2017) who tested the effect of iron sulfide minerals on each of the citric acid cycle reactions. They observed a significant catalytic effect of ferrous sulfide and peroxydisulfate on most of the reactions, which is consistent with the idea that similar catalyzed reactions could occur in the prebiotic environment.

Redox Reactions: Sources and Sinks

Another approach is based on the fact that most of life's metabolic energy other than photosynthesis involves the downhill transport of electrons available in reduced compounds to more oxidized compounds such as molecular oxygen. The reactions are usually shown on a scale with units expressed as voltages. For instance, the potential energy made available by coupling electron transport to proton transport across a membrane is equivalent to 0.2 volts, which is sufficient to drive ATP synthesis in mitochondria and chloroplasts.

Figure 7.2 shows how individual metabolic steps can be related to energy made available from catalyzed redox reactions. In this figure, energy content of reactants is expressed in units ranging from negative to positive millivolts. By convention, a redox reaction can occur spontaneously with release of energy when a compound higher on the scale loses electrons to become a product lower on the scale. The voltage difference

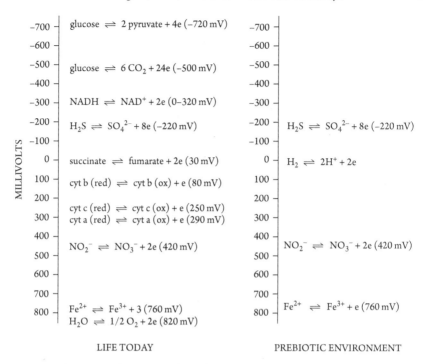

$$glucose \rightleftharpoons 2\,pyruvate + 4e \,(-720\,mV)$$

$$glucose \rightleftharpoons 6\,CO_2 + 24e \,(-500\,mV)$$

$$NADH \rightleftharpoons NAD^+ + 2e \,(0\text{–}320\,mV)$$

$$H_2S \rightleftharpoons SO_4{}^{2-} + 8e \,(-220\,mV)$$

$$succinate \rightleftharpoons fumarate + 2e \,(30\,mV)$$

$$cyt\,b\,(red) \rightleftharpoons cyt\,b\,(ox) + e \,(80\,mV)$$

$$cyt\,c\,(red) \rightleftharpoons cyt\,c\,(ox) + e \,(250\,mV)$$
$$cyt\,a\,(red) \rightleftharpoons cyt\,a\,(ox) + e \,(290\,mV)$$

$$NO_2{}^- \rightleftharpoons NO_3{}^- + 2e \,(420\,mV)$$

$$Fe^{2+} \rightleftharpoons Fe^{3+} + 3 \,(760\,mV)$$
$$H_2O \rightleftharpoons 1/2\,O_2 + 2e \,(820\,mV)$$

$$H_2S \rightleftharpoons SO_4{}^{2-} + 8e \,(-220\,mV)$$

$$H_2 \rightleftharpoons 2H^+ + 2e$$

$$NO_2{}^- \rightleftharpoons NO_3{}^- + 2e \,(420\,mV)$$

$$Fe^{2+} \rightleftharpoons Fe^{3+} + e \,(760\,mV)$$

LIFE TODAY   PREBIOTIC ENVIRONMENT

Figure 7.2. The chemical reactions of metabolism expressed in terms of electrochemistry. Those used by life today are on the left, and possible sources of metabolic energy in prebiotic conditions are on the right.

between glucose and oxygen is approximately 1400 millivolts, or 1.4 volts, and this energy is released in mitochondria when the electrons in glucose are made available as pyruvate by glycolysis, then transported through a series of enzyme-catalyzed reactions to molecular oxygen. The energy is captured in three molecules of ATP synthesized for each pair of electrons traveling from pyruvate to oxygen.

Premetabolic Chemical Reactions in Hydrothermal Conditions

The reactions shown in Figure 7.2 and their related energy content are basic knowledge in chemistry and can be used to understand how early forms of life might capture metabolic energy from the environment. When hydrothermal vents were discovered, Jack Corliss and John Baross suggested that life might have originated in hydrothermal vents, a proposal that inspired an extensive series of publications by Michael Russell, William Martin, Nick Lane and their students. Martin and Russell (2007) and Sojo et al. (2016) described how reduction of CO_2 in vent conditions might be a source of reduced carbon compounds to support a primitive metabolism. This idea was recently tested in Lane's laboratory in a simulation of hydrothermal vents that will be described in Chapter 8.

The main ideas were summarized by Lane (2015), and depicted in Plate 7.2 (Lane and Martin, 2012). The authors propose that the vent minerals contain iron–nickel sulfides that catalyze a reaction in which molecular hydrogen reduces CO_2 to thioacetate by linking it to thiomethane in the solution (top panel). The thioacetate becomes phosphorylated and the activated compound can be used as a source of high energy phosphate.

The middle panel shows the compartment at a later stage of evolution. A pH gradient is present across the mineral membrane because vent fluids are alkaline and seawater is acidic. A lipid membrane now seals the inner surface of the compartment so that proton gradients can be maintained. Protein synthesis has been established, and one of the proteins is an enzyme similar to ATP synthase that can use the chemiosmotic energy of the proton gradient to synthesize ATP.

The bottom panel shows a proton–sodium exchanger, embedded in the bilayer, that can exchange internal protons for sodium ions. ATP synthesis now has a second source of energy in the form of sodium ions moving outward through the synthase

This is one of the most detailed proposals for how metabolism could begin and is therefore worth summarizing at three levels to keep facts and assumptions separate from conjecture. I am not using the word conjecture in a pejorative sense, and some of the ideas I will present in Chapter 10 would be classified as conjecture.

Facts (from Chapter 3):

- Hydrothermal vents occur along MORs as black smokers driven by magmatic heat. The emitted fluids are extremely hot and rich in dissolved sulfides that precipitate as mineral chimneys upon encountering cool seawater.
- A second type of hydrothermal vent is driven by serpentinization. The fluids are at lower temperatures, very alkaline, and rich in dissolved hydrogen. Their dissolved components precipitate as chimneys of carbonate minerals.
- The chimneys are porous and have apparent compartments ~100 micrometers in diameter.
- Microbial life has found a niche in both types of vents.

Assumptions and conjectures not tested by experiments or observation:

- The vent minerals can catalyze the reduction of CO_2 by the dissolved hydrogen.
- A substantial pH gradient develops between the alkaline fluid and relatively acidic seawater.
- The porous minerals compartments can concentrate potential reactants and maintain products within their volume.
- The reduced carbon compounds can act as a substrate to initiate a primitive version of metabolism.
- A lipid film seals the otherwise porous mineral membranes so that pH gradients can be maintained.
- The pH gradients can serve as a source of chemiosmotic energy coupled to pyrophosphate bond synthesis.

- Catalytic functions similar to those of ferredoxin can be performed by metal sulfide minerals.
- An ATP synthase embedded in the mineral membrane can synthesize ATP from ADP and dissolved phosphate.

As is proper in scientific research, recent papers from other experts express skepticism about whether CO_2 can actually be reduced under vent conditions, as well as the conclusions that follow from that assumption. For instance, Jackson (2016) makes the point that mineral membranes are much too thick to function as a chemiosmotic membrane because chemiosmosis requires a thin membrane to harvest the energy of a pH gradient. An analogy that may help clarify this point is to imagine a turbine and generator capturing the energy of a 10-meter waterfall over a cliff (a "sharp" gradient) and then putting the same generator into a river that falls the same distance—10 meters—but over a distance of one kilometer (analogous to a thick membrane). Virtually no energy can be captured from the slow-moving river.

Phosphate's Role in Primitive Metabolism?

ATP is an extraordinary molecule that has permeated all life as an energy currency, yet the origin of this particular molecule is a major gap in our understanding. To the eye of a biochemist, ATP has a certain beauty when illustrated as in Plate 7.3. Cells use ATP to gather energy from a source such as photosynthesis or respiration and then deliver it to energy-requiring reactions in the rest of the cytoplasm. The chemical-energy content of ATP is present in the pyrophosphate bond that links its second and third phosphate groups. This is called a high-energy bond because of its relatively large energy content, expressed as kilocalories per mole of ATP (kcal/mol) or as kilojoules per mole (kJ/mol) in international units. To give a sense of the amount of energy available in the bond, the units were originally measured as calories, with units defined as the amount of heat that raises the temperature of a gram (approximately one cubic centimeter) of water by one degree Celsius. (This is simplified, but the technical definition is not necessary here.) A kilocalorie will therefore raise the temperature of a liter of water (one thousand grams) by one degree C. If one mole of ATP (507 grams) is dissolved in one liter of water and allowed to hydrolyze, the temperature of the water would increase by approximately 7 degrees C. This energy, by the way, is a chemist's version which is measured under carefully defined conditions of temperature and concentration. In a living cell, the conditions are quite different and ATP makes a greater amount of energy available, closer to 10 kcal per mole.

ATP hydrolysis is the primary source of the heat used by mammals and birds to maintain their body temperature at a fixed point above that of the environment. Here is the hydrolysis reaction under standard conditions:

$$\text{ATP} + \text{H}_2\text{O} \Leftrightarrow \text{ADP} + \text{phosphate} + \text{heat} \ (7.6\,\text{kcal/mol})$$

Pyrophosphate bonds are referred to as anhydrides because they are synthesized when the elements of water are removed from between two acidic compounds rather than the acid and an alcohol of ester bonds. The chemical energy of anhydride bonds can be released by energetically downhill group transfer reactions of the phosphate group to other molecules, an enzyme-catalyzed activating process called phosphorylation. The second molecule gains chemical energy and can in turn undergo reactions that otherwise will not occur. One example of the activating effect is phosphorylation of glucose catalyzed by hexokinase. In the absence of phosphorylation, glucose would be unable to enter the glycolysis pathway, and its energy content could not be released to the cell.

$$\text{ATP} + \text{glucose} \Leftrightarrow \text{glucose-6 phosphate} + \text{ADP}$$

The energy of ATP is also used to drive sodium, potassium, and calcium ion transport across cell membranes to produce concentration gradients essential for life. Significantly, hydrogen ions (protons) are also transported by ATP-dependent enzymes, and the importance of this will be discussed later.

But how could ATP first be incorporated into a primitive metabolism? Could there be a simpler molecule that also contains energy in anhydride bonds? In fact, if a solution of ordinary phosphate is dried and heated to temperatures above 200° C, small yields of pyrophosphate (abbreviated PPi with the "i" signifying inorganic) are produced. Pyrophosphate is composed of two phosphates linked by an anhydride bond that has a little less energy content than ATP but still sufficient to serve as a source of phosphate in certain reactions:

$$\text{PPi} + \text{H}_2\text{O} \Leftrightarrow 2\,\text{PO}_4 + 5.3\,\text{kcal/mol}$$

Holm and Baltscheffsky (2011) argued that pyrophosphate might serve as a primary energy source for primitive life and cited two facts that support this idea. First, small quantities of pyrophosphate-containing minerals have been discovered in quarries, so it may have been available on the early Earth. Second, and more important, the coupling membranes of a photosynthetic bacterial species called *Rhodospirrilum rubrum* synthesize pyrophosphate rather than ATP, using the pyrophosphate as an energy source much as other organisms use ATP. Perhaps this reaction is a kind of molecular fossil left over from the earliest metabolic pathways that incorporate phosphate as an energy currency.

Sulfur Chemistry and Metabolism

The aroma of both hydrogen sulfide and sulfur dioxide is noticeable in the atmosphere near active volcanoes today, and elemental sulfur often forms yellow crusts around volcanic vents. (An older word for sulfur is brimstone, derived from old English words meaning burning stone because elemental sulfur actually burns with a blue flame when ignited.) Hydrogen sulfide and metal sulfides are also abundant components in the fluids and chimneys of hydrothermal vents. Microbial life in the vents has developed a metabolism that can harvest the energy of hydrogen sulfide by using it as a source of reducing power, meaning that it can donate electrons to compounds farther down on the redox scale shown in Figure 7.2. This has inspired several proposals that sulfur chemistry might also be involved in primitive metabolic pathways of early life. Because the thiol ester has a greater energy content (7 kcal/mol) than ordinary esters (3 kcal/mol), Christian DeDuve (1998) suggested that high-energy thioesters might have provided the energy for metabolism before ATP synthesis evolved. Thioesters are synthesized when a water molecule is lost during the reaction of an organic acid and a thiol:

$$R - COOH + R' - SH \Leftrightarrow R - CO - SR' + H_2O$$

Wächtershäuser (1990, 1998) visualized a more extensive series of reactions involving sulfur chemistry that use reducing power generated when hydrogen sulfide reacts with dissolved ferrous iron to produce the insoluble iron sulfide mineral called pyrite. It was proposed that the energy released by the reaction could drive energetically uphill chemical reactions on the surface of the pyrite, which otherwise cannot take place in solution. This idea was experimentally tested in a simulation that will be described in Chapter 8.

Chemical Energy of Organic Substrates: Carbohydrates

As mentioned in Chapter 4, simple carbohydrates are abundant products of the formose reaction, so given a source of formaldehyde on the early Earth it is reasonable to assume that they were being continuously synthesized in the mixture of solutes in hydrothermal water. Carbohydrates contain both chemical energy and reactivity that is largely absent in other biologically relevant organic compounds such as amino acids and may have played a significant role in establishing the first metabolic pathways. This property has been extensively explored by Art Weber (2000) who points to the example of anaerobic fermentation reactions, noting that as the internal energy of glucose is released in glycolysis, ATP is synthesized by a process that does not require the complexity of membranes, electron transport enzymes, proton transport, and an ATP synthase. This is the simplest known metabolic reaction that generates the anhydride bond of ATP. Carbohydrates can also undergo spontaneous reactions with themselves and with solutes such as ammonia to produce a variety of other biologically relevant compounds like amino acids. In other words, given the occurrence of the formose reaction and the presence of amines, the reactions of carbohydrates could generate most of the simple molecules required for metabolic pathways to begin.

One such reaction may even provide a clue to the way that homochirality was initiated in early forms of life. Pizzarello and Weber (2004) investigated a reaction in which glycolaldehyde spontaneously reacted with itself to form the 4-carbon sugars erythrose and threose. The products would normally be racemic, in other words, an equal mixture of D and L sugars. However, they found that the presence of an L-amino acid in the mixture led to a remarkable enantiomeric enhancement of the D form of the sugars that were synthesized! This is significant for two reasons. First, D sugars and

L amino acids are the chiral species used by all life on Earth. Second, in other studies Pizzarello et al. (2008) reported that the amino acids of carbonaceous meteorites had an unexpected L enhancement. It is interesting to speculate that the homochirality of life today may have been initiated by prebiotic chemical reactions in which the excess L amino acids delivered by meteorites tipped the balance toward D sugars by influencing prebiotic reactions related to carbohydrate synthesis.

Primitive Metabolism must Function in Compartments

Most of the reactions just described occur in aqueous solutions, but this book emphasizes that the origin of life was in fact the origin of cellular life, so it follows that the reactions of a primitive metabolism at some point in early evolution became established in membranous compartments. If we accept this premise, we are immediately faced with multiple questions that must be addressed:

1. How did potential reactants cross the barrier of a membrane to enter the compartment?
2. What source of energy was available to drive metabolism in the compartment?
3. What catalysts became incorporated into metabolic pathways?
4. How did metabolic catalysts evolve in response to environmental stresses?
5. How did feedback loops emerge to regulate primitive metabolism?

There have been very few studies that attempt to answer these questions, so what follows would be characterized as conjecture, though it is informed conjecture, because it can be tested experimentally.

The first point to make is that most of the reactions of metabolism do in fact occur in aqueous solutions, and biochemists are used to carrying out reactions in a few milliliters of solution in a glass test tube. They are properly skeptical of the idea that reactions could equally well occur in microscopic volumes, but that is what we are proposing. In other words, lipid vesicles with diameters of a few micrometers are simply microscopic test tubes, and reactions that work in milliliter volumes of a test tube will work equally well in the nanoliter volume of a vesicle one micrometer in diameter. This point will become clear in the description that follows.

In my estimation, the most plausible original encapsulation process is a cycle of dehydration followed by rehydration. This was established experimentally years ago (Deamer and Barchfeld, 1982) and is one of the primary steps in the pathway to cellular life that will be described in Chapters 9 and 10. Given that potential reactants can become encapsulated, it is not difficult to imagine a process by which proton gradients can be produced across the boundary membranes of vesicular compartments. The deep interior of the early Earth was a source of reducing power in the form of gases such as hydrogen and methane, produced by reactions of water with minerals at high temperatures and pressures (Fig. 7.3). The dissolved gases react with mineral components to form reduced compounds like FeS_2. A redox boundary would exist between the reduced

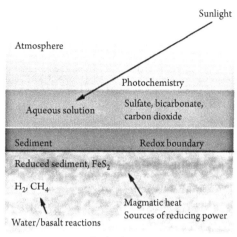

Figure 7.3. Sources of redox energy in the prebiotic environment. Adapted from Fishbaugh et al. 2007.

compounds in the sediment and the relatively oxidized compounds like sulfate, bicarbonate, and dissolved CO_2 in the atmosphere.

Given that redox boundaries existed on the early Earth, is there an alternative to hydrothermal vents as a means for primitive life to take advantage of the available energy? In fact, chemical energy is generated in a closed, membrane-bounded compartment if an electron donor at a higher redox state is present in the external medium and an acceptor at a lower redox state is inside. If a carrier that can partition into the lipid bilayer is present, it is able to pick up electrons from the donor and transport them across the membrane to be delivered to the acceptor (Fig. 7.4). Shortly after liposomes were first established as a model membrane system, we demonstrated such a system in which encapsulated ferricyanide was an acceptor and ascorbic acid was a donor (Deamer et al., 1972). Phenazine methosulfate was added as an obligatory proton acceptor that added protons to its structure when it is reduced. The carrier transported protons and electrons across the membrane into the inner volume and the protons were released when the electrons were delivered to the ferricyanide. The interior became acidic, and a very large gradient of several pH units developed across the membrane.

What electron donors, acceptors, and carriers are plausible components of the prebiotic environment? It is unlikely that either sulfate or CO_2 could be direct acceptors of electrons due to their low reactivity. Iron, either as ions in solution or in the form of iron compounds, readily accepts electrons during redox reactions. Iron also plays a major role in electron transport today, so iron compounds are a reasonable choice to serve as electron acceptors.

For an electron donor, it is necessary to choose a compound that cannot easily cross membranes because the electrons and protons must be picked up from the external medium in order to undergo transport across the membrane. This excludes hydrogen gas, hydrogen sulfide, and methane. A more likely donor would be reduced sulfur

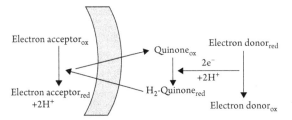

Figure 7.4. Closed membrane compartments in which electron donors, acceptors, and carriers can establish an electron transfer reaction across the membrane, which produces electrochemical proton gradients.

compounds which can donate electrons in redox reactions taking place in aqueous phases. It is interesting that sulfur in the form of iron-sulfur complexes is an essential component of electron transport systems today. Presumably its role in bioenergetic processes emerged because of its abundance in the Earth's crust and the multiple oxidation states that sulfur compounds can assume.

The last component of a primitive electron and proton transport system concerns small molecules that can act as carriers. Life today uses quinones to transport both electrons and protons across membranes. Examples include ubiquinone in mitochondria and bacteria and plastoquinone in chloroplasts. The incorporation of quinones into the chemiosmotic systems of microorganisms is a significant clue to the mechanisms of electron and proton transport in early cellular life, because quinones such as anthracenedione are present in carbonaceous meteorites. From this comes a reasonable assumption that the prebiotic organic mixture included polycyclic aromatic derivatives such as quinones that could act as electron transporters across membranes and thereby develop a proton gradient to be used as a source of chemiosmotic energy. This coupled reaction is currently being tested in my laboratory using polycyclic compounds extracted from the Murchison meteorite.

Conclusions and Open Questions

Although we have a detailed understanding of the energy sources and incredible complexity of metabolic pathways in living cells today, we know very little about even the simplest reactions that might have been available for primitive life. There are guesses, of course, and abundant conjecture, but this remains a virtually unexplored question for experimentalists, with much to be discovered. My advice to a young investigator interested in this problem would be to begin with determining how membranous compartments could have captured light energy using a plausible pigment system and how the pigment could have generated a proton gradient across membranes. A second question open for discovery is how activated phosphate could be generated in plausible prebiotic conditions then encapsulated along with an obvious potential substrate such as glyceraldehyde and whether phosphorylation can then spontaneously occur. Phosphorylation is an essential driver in metabolism today, and there must have been a primitive version, but we have no idea how this reaction could have emerged and been incorporated into the first forms of cellular life.

The last list of bullet points below summarizes the facts and assumptions that are the foundation for the argument that will be proposed in the next two chapters.

- Small organic molecules from sources such as those described in Chapter 4 are present as dilute solutions in the early ocean but accumulate in more concentrated form in acidic hydrothermal pools on volcanic land masses.
- Chemically active small molecules in solution include cyanide and formaldehyde. Typical bilayer lipid membranes are relatively permeable to both species.
- Phosphate dissolved from minerals by acidic water is in the solution. Membranes are relatively impermeable to phosphate and other ionic solutes.
- Molecular oxygen is absent in the early atmosphere but there is abundant CO_2.
- Reducing power is available in the form of dissolved hydrogen gas and hydrogen sulfide as donors, but if energy is to be made available from redox reactions there must also be an oxidant that can accept electrons.
- Metal sulfides with the potential to catalyze chemical reactions are present in the form of exposed mineral surfaces in freshwater pools.
- Organic solutes can undergo a variety of chemical reactions driven by photochemistry, by redox reactions, and by condensation reactions to produce more complex molecules, including polymers.
- Amphiphilic compounds are present and assemble into membranous compartments in the form of vesicles.
- If complex mixtures of organic compounds undergo reactions driven by local energy sources, increasingly complex molecular systems emerge in a steady state far from equilibrium.
- The trend toward more complex systems is subject to the laws of thermodynamics and kinetics, so the steady state is balanced between synthesis of complex compounds and decomposition reactions, just as it is in life today.

References

Cody GD, Boctor NZ, Hazen RM, Brandes JA, Morowitz HJ, Yoder HS (2006) Geochemical roots of autotrophic carbon fixation: Hydrothermal experiments in the system citric acid, H_2O-($\pm$FeS)–($\pm$NiS). *Geochim Cosmochim Acta* 65, 3557–3576.

Cooper G, Reed C, Nguyen D, Carter M, Wang Y (2011) Detection and formation scenario of citric acid, pyruvic acid, and other possible metabolism precursors in carbonaceous meteorites. *Proc Natl Aca Sci USA* 108, 14015–14020.

Deamer DW, Barchfeld GL (1982) Encapsulation of macromolecules by lipid vesicles under simulated prebiotic conditions. *J Mol Evol* 18, 203-206.

Deamer DW, Prince RC, Crofts AR (1972) The response of fluorescent amines to pH gradients across liposome membranes. *Biochim Biophys Acta* 274, 323–355.

DeDuve C (1998) Clues from present-day biology: The thioester world. In A Brack (Ed.) *The Molecular Origins of Life* (pp. 219–236). Cambridge: Cambridge University Press.

Fishbaugh KE, Lognonné P, Korablev O, Des Marais DJ, Raulin F (2007) *Geology and Habitability of Terrestrial Planets*. New York: Springer.

Holm NG, Baltscheffsky H (2011) Links between hydrothermal environments, pyrophosphate, Na+, and early evolution. *Orig Life Evol Biosph* 41, 483–493.

Jackson BJ (2016) Natural pH gradients in hydrothermal alkali vents were unlikely to have played a role in the origin of life. *J Mol Evol* 83, 1–11.

Keller MA, Kampjut D, Harrison SA, Ralser M (2017) Sulfate radicals enable a non-enzymatic Krebs cycle precursor. *Nature Ecology & Evolution*. doi: 10.1038/s41559-017-0083.

Lane N (2015) *The Vital Question: Energy, Evolution, and the Origins of Complex Life*. New York: W.W. Norton.

Lane N, Martin WF (2012) The origin of membrane bioenergetics. *Cell* 151, 1406–1416.

Martin W, Russell MJ (2007) On the origin of biochemistry at an alkaline hydrothermal vent. *Phil Trans R Soc Lond B* 362, 1887–1925.

Morowitz HJ (1992) *The Beginnings of Cellular Life*. New Haven, CT: Yale University Press.

Morowitz HJ, Kostelnik JD, Yang J, Cody GD (2000) The origin of intermediary metabolism. *Proc Natl Acad Sci USA* 97, 7704–7708.

Pizzarello S, Weber AL (2004) Prebiotic amino acids as asymmetric catalysts. *Science* 303, 1551–

Pizzarello S, Huang Y, Alexandre MR (2008) Molecular asymmetry in extraterrestrial chemistry: Insights from a pristine meteorite. *Proc Natl Acad Sci USA* 105, 3700–3704.

Sojo V, Herschy B, Whicher A, Camprubi E, Lane N (2016) The origin of life in alkaline hydrothermal vents. *Astrobiology* 16, 181–197.

Wächtershäuser G (1990) Evolution of the first metabolic cycles. *Proc Natl Acad Sci USA* 87, 200–204.

Wächtershäuser G (1998) Origin of life in an iron-sulfur world. In A Brack (Ed.) *The Molecular Origins of Life* (pp. 206–218). Cambridge: Cambridge University Press.

Weber AL (2000) Sugars as the optimal biosynthetic carbon substrate of aqueous life throughout the universe. *Orig Life Evol Biosph* 30, 33–43.

Testing Alternative Hypotheses

Simulating the Prebiotic Environment

> Strong inference consists of applying the following steps to every problem in science, formally and explicitly and regularly: 1) Devising alternative hypotheses; 2) Devising a crucial experiment (or several of them), with alternative possible outcomes, each of which will, as nearly as possible, exclude one or more of the hypotheses; 3) Carrying out the experiment so as to get a clean result.
>
> John Platt, 1964

Overview and Questions to Be Addressed

Narratives related to the origin of life all incorporate the assumption that the first living microorganisms emerged in a sterile planetary surface after the ocean condensed over 4 billion years ago. This means that even though global conditions can be deduced from our growing understanding of planetary science and the early solar system, laboratory simulations of localized prebiotic sites remain the only way to guide virtually all of the attempts to reproduce the chemical and physical processes by which life could emerge. Because simulations are used to test the hypotheses described in this book, it seems essential to review representative examples and consider their strengths and limitations.

Laboratory simulations of the prebiotic environment incorporate specific sets of factors chosen to reflect what we can deduce about the atmosphere, lithosphere, and hydrosphere of the early Earth and Mars. This chapter attempts to sort out the main factors that are incorporated into simulations. These can be conveniently divided into the drivers—physical and chemical processes—and the emergent complexity that is a product of the drivers. In order to describe a given system, it is helpful to first provide abbreviations for the component factors that define the degree of complexity, then compare that complexity to the complexity that seems to be required for life to begin, as is done in the next section.

Questions to be addressed:

- What factors and assumptions provide a foundation for a given simulation?
- How do the factors interact to produce increasing complexity?

- How can the factors guide the design of experimental approaches?
- What combination of factors would be an adequate simulation of the prebiotic condition?

A Nomenclature to Describe Prebiotic Simulations

Laboratory investigations related to origin of life research often begin with mixtures of simple organic molecules assumed to be available on the prebiotic Earth. These are then allowed to undergo reactions driven by a source of energy, either impinging on the system such as ultraviolet light or electrical discharge or contained within chemically activated compounds that can undergo spontaneous reactions. The mixture typically becomes more complex in some interesting way, such as the synthesis of polymers described in earlier chapters. The results are then taken as clues to the pathway by which the products can perform functions such as encapsulation into membranous compartments, energy capture, catalyzed growth, replication, metabolism, and ultimately evolution.

At first glance, the various simulations seem to have little in common, making it difficult to judge how well they represent prebiotic conditions. However, it is possible to sort out a limited set of factors that are either present or absent in any given simulation. Some of the simulations are quite simple and only incorporate one or two factors, but it is obvious that such systems lack the complexity sufficient to support the emergence of life. How many factors would be sufficient? That is the question addressed in this chapter, and the list that follows includes eighteen factors that can contribute to the needed complexity. For convenience, each factor is followed by an abbreviation that will be used to evaluate examples published in the literature, with a particular emphasis on simulations of hydrothermal vents and fields.

Physical processes:

self-assembly of molecular structures such as by adsorption onto surfaces or templates, base pairing, folded polymers, or boundary membranes (As)
cycles such as light–dark or hydration–dehydration (Cy)
encapsulation of reaction mixtures and products derived from self-assembly such as the formation of lipid vesicles (En)
transport to accumulate nutrients from the external environment (Tr)
concentration of reactants (Cn)—an active process such as adsorption or evaporation that concentrates otherwise dilute reactants
heating (He)—in chemical laboratories a mixture of reactants is often heated to add activation energy and make a reaction go faster, and some of the simulations are heated for this purpose

Chemical processes:

chemical energy is either present in the reactants or is added by impinging energy sources (Ce) *condensation reactions* lead to polymerization when ester or peptide bonds are synthesized (Cd)

metabolism chemically alters nutrients to be used for energy and growth (Me)

kinetic traps in which the rate of synthesis exceeds a back reaction so that products accumulate in a steady state (Kn)

Emergent processes:

catalytic and autocatalytic activity, either of environmental components such as metals, or emerging in products of reactions (Ca)

hill climbing, in which a given reaction mixture becomes more complex as products accumulate and begin to interact (Cl)

feedback circuits that regulate reaction rates and products (Fe)

combinatorial selection of systems incorporating molecular systems that vary in composition or amounts (Cm)

coding of a sequence of monomers in one polymer that can direct the monomer sequence during synthesis of a second polymer (Co)

mutations occur when errors are made during coding (Mu)

selection takes place when populations of molecular systems are exposed to conditions that allow only some of the systems to survive (Se)

evolution begins when certain molecular systems are either more efficient at capturing and utilizing resources or survive an environmental stress that disrupts less robust systems (Ev)

Each of these eighteen factors can affect reactants and their products as they undergo cyclic iterations and are the minimal number required for a sufficiently complex simulation in which an input mixture of simple reactants yields an increasingly complex mixture of products related to the origin of life. They can be expressed as a function of time, and we will use the following scheme to describe various simulations:

$$\text{Reactants} \Rightarrow \left(\text{As, Cy, En, Tr, Cn, He, Ce, Cd, Me, Kn, Ca, Cl, Fe, Cm, Co, Mu, Se, Ev}\right) \text{ time}$$
$$\Rightarrow \textbf{Increasing complexity}$$

The notation should not be considered to represent typical chemical reactions in which reactants proceed to equilibrium. Instead, the notation reflects reactants and products interacting in a steady state maintained by a source of energy. It is significant that the emergence of one or more products having catalytic activity will shift one steady state to an alternative steady state. Simulations incorporating several lower level factors support the emergence of higher level factors so that a natural hierarchy is established.

The products of a given simulation are highly variable and usually unpredictable, but they have two general properties. They are more complex than the starting components of the system (Adami et al. 2000) and they are in some way related to a property or function of life. Examples include amino acids synthesized from gas mixtures, fatty acid products of FT reactions, sugar products of the formose reaction, and polymers produced by condensation reactions. The products can have a second emergent property such as self-assembly into compartments or catalytic activity if they are polymers. The overall goal of this exercise is to sort reactants, conditions, and products into categories that summarize specific properties of a given simulation. This allows the simulation to be evaluated in terms of how closely it approaches a complexity sufficient to support the origin of life.

Evaluating Examples of Simulations

We will now describe several simulated prebiotic conditions from the literature to illustrate how application of the nomenclature can characterize simulations and test them for sufficient complexity.

Reactants → (Ce, Kn, He) time → Products

Miller (1953) exposed a mixture of hydrogen, methane, ammonia, and water vapor at atmospheric pressure to a spark discharge for several days (He) (Plate 8.1). The discharge (Ce) chemically activated the gasses to compounds such as HCN and HCHO, which then reacted according to the Strecker reaction to form a variety of amino acids and other compounds that accumulated in a kinetic trap (Kn). A significant point is that no one would expect life to begin in the conditions of the Miller experiment. The simulation is insufficiently complex.

Reactants → (Ce, He) time → Products

Given Miller's results, Sidney Fox (1964) assumed that solutions of amino acids would be dried and heated on prebiotic volcanic mineral surfaces and tested whether heat could drive polymerization. In a typical experiment, a mixture of amino acids was heated (He) for several hours to temperatures (160° C) at which they melted and polymerized (Ce). Furthermore, if the polymer was boiled in water and cooled, it formed large numbers of microscopic structures Fox called proteinoid microspheres. Fox became certain that he had discovered the path to primitive life, but other scientists were skeptical because most of the linking bonds were undefined, and there was no obvious way for proteinoids to function as replicating systems and evolve.

Reactants → (Ce, Cy, He) time → Products

Rodriguez-Garcia et al. (2015) employed an automated cycling device to expose glycine solutions containing 0.25 M NaCl to multiple wet–dry cycles. This resulted in short oligopeptides ranging from dimers to dodedcamers.

Reactants → (Ce, Cy, En, Kn, He) time → Products

Furuuchi et al. (2005) tested a simulation of hydrothermal vents in which seawater circulated in cycles (Cy) between hot, high pressure conditions followed by quenching in cold, low pressure conditions. Glycine (100 mM) and fatty acids were added to the seawater, and it was found that oligoglycine peptides up to hexamers were formed. In this case, the energy source was most likely the fact that in hot, pressurized water the energy barrier for peptide bond formation is significantly reduced. The glycine therefore enters a steady state with oligoglycine, and the peptides accumulate in a kinetic trap after quenching (Kn). Although the presence of lipids promoted peptide bond synthesis, it is unclear whether this was a catalytic effect or related to encapsulation (En).

$$Reactants \rightarrow (Ce, En, Cn) \; time \rightarrow Products$$

Sousa et al. (2013) pointed out that hydrothermal vent minerals are porous and have multiple microscopic channels in which solutes can be concentrated (Cn). Furthermore, pH gradients are proposed to be maintained in alkaline vents across mineral boundaries of the channels and compartments, providing a form of chemiosmotic energy that could drive chemical reactions. This represents the Ce term in the formulation above, and the mineral compartments are reflected in the En term. This is an untested hypothesis. So far there has been no convincing laboratory demonstration that mineral chemiosmosis can drive synthesis of products related to the origin of life.

$$Reactants \rightarrow (Ce, Cn, He) \; time \rightarrow Products$$

Huber and Wächtershäuser (1997) heated an aqueous dispersion of nickel and iron sulfide in the presence of CO and methanethiol, producing the thioester $CH3$-CO-SCH_3 as a product, which upon hydrolysis, formed acetic acid. The authors considered this to be a "primordial initiation reaction for a chemoautotrophic origin of life."

$$Reactants \rightarrow (Ce, Cn, He) \; time \rightarrow Products$$

McCollom et al. (1999) heated oxalic acid to 175 °C for 2–3 days in a pressure bomb. The heat energy (Ce) caused the oxalic acid to dissociate into hydrogen gas and CO which then reacted with the surfaces of the stainless steel enclosure by a FT process to produce long-chain hydrocarbons and derivatives such as alcohols and monocarboxylic acids ranging in length from 2 to 35 carbons.

$$Reactants \rightarrow (Ca, As, En, Se) \; time \rightarrow Products$$

Adamala and Szostak (2013) established a system of self-assembled (As) fatty acid vesicles, some with (En) and others without a serine-histidine dipeptide (Ca) that catalyzes the synthesis of another relatively hydrophobic dipeptide. When the two populations were mixed, vesicles containing the catalytic dipeptide grew at the expense of those lacking the dipeptide. The hydrophobic product bound to membranes and reduced the energy barrier for fatty acid assimilation from the solution, thereby promoting growth. This result represents one of the simplest examples of selection (Se) in which one population of vesicles competes with another.

Reactants → (Ce, Cn, Cd, Kn, Ca, As) time → Products

Ferris and coworkers (2006) incubated chemically activated ribonucleotides (Ce) for several days with montmorillonite clay. The monomers adsorbed to specific sites on the surface (Cn), followed by condensation (Cd) into oligonucleotide products that accumulated in a kinetic trap (Kn). The clay can therefore be considered to catalyze the reaction (Ca) when the monomers assembled on the clay surfaces (As).

Reactants → (Ce, Cd, Kn, Ca, As, Co) time → Products

Inoue and Orgel (1983) allowed chemically activated ribonucleotides (Ce) to interact with RNA templates in the cold for several days. The templates had specific base sequences that acted as a catalyst (Ca) by assembling and concentrating monomers along their strands by complementary base pairing (As, Cn). The products were complementary to the coded sequences (Co) and accumulated in a kinetic trap (Kn).

Reactants → (Ce, Cy, Cd, Kn, Ca, As, Co, Mu, Ev) time → Products

Lincoln and Joyce (2009) used in vitro evolution and mutagenic polymerase chain reaction (Mu) and selection (Se) to establish pairs of cross-replicating ribozymes. The ligation reaction was driven by pyrophosphate bond energy (Ce) and cycling was performed by serial dilution into fresh media (Cy). Under these conditions cross-replication could go on indefinitely by self-assembly of the substrates (As) according to coded sequences (Co). Spontaneous mutations allowed the system to evolve new polymers with novel functions (Ev).

Reactants → (Ce, Cy, Cd, Kn, Cl, Ca, As, Cm, Co, Se, Ev) time → Products

Bartel and Szostak (1993) exposed random sequence RNA to multiple cycles (Cy) of ligation and amplification. The amplification required enzymatic catalysis (Ca) and energy was provided by the pyrophosphate bonds in nucleoside triphosphates (Ce) that assembled on templates (Co) during amplification (As). Selection (Se) occurred during each cycle, products were maintained in a kinetic trap (Kn), and hill climbing (Cl) proceeded as functional ribozymes were selected from random sequences of RNA. The result was that novel species of RNA with ribozyme activity were generated by an evolutionary process (Ev).

Reactants → (Ce, Cd, Kn, Ca, As, En, Co, Tr) time → Products

Noireaux and Libchaber (2004) encapsulated (En) a cytosolic lysate of E. coli bacteria in lipid vesicles along with two plasmids, one for green fluorescent protein (GFP) and one for hemolysin. They demonstrated synthesis of GFP in the vesicles and transport (Tr) of substrates added to the external medium through hemolysin pores. More recently, Ichihashi et al. (2013) encapsulated ribosomes with a complete translation system in lipid vesicles. These systems are not directly related to the origin of life but instead would be defined as reconstitution. Although the systems come closer to sufficient complexity, they lack cycles (Cy), metabolism (Me), feedback (Fe), selection (Se), mutations (Mu), and evolution (Ev).

We will now compare these examples to simulations of hydrothermal vents and hydrothermal fields that are directly related to the alternative hypotheses described in this book.

Reactants → (Ce, Ca, Cm, Me) time → Products

This is a simulation of the hydrothermal vents described in Chapter 3. Herschy et al. (2014) constructed a reaction vessel in which a high pH fluid was slowly mixed with an acidic fluid to simulate the pH gradients assumed to exist across the mineral membranes of alkaline hydrothermal vents. The atmosphere was a 98:2 mixture of nitrogen and hydrogen gas. The acidic fluid (pH 5) was a dilute mixture of $FeCl_2$, $NaHCO_3$, and $NiCl_2$, and the alkaline fluid (pH 11) was composed of K_2HPO_4, $Na_2Si_3O_7$, and Na_2S. To simulate the mineral deposition associated with alkaline vents, the alkaline fluid was slowly injected (at a rate of 10–120 microliters per hour) into a larger volume of the acidic fluid in the reaction vessel (Plate 8.2). Microscopic precipitates of ferrous silicates and phosphates formed, which appeared to resemble hollow tubes when examined by scanning electron microscopy. The authors proposed that proton gradients across the mineral membranes of the tubes could drive the reduction of CO_2. Because CO_2 was present in the mixture from the equilibrium of bicarbonate and CO_2 at pH 5, measuring expected products of CO_2 reduction could test this proposal. After several hours, traces of formic acid could be measured at 50 μM concentrations, and in some of the experimental runs, trace amounts (100 nM) of formaldehyde were also detected.

Although the authors proposed that these results are consistent with the idea that energy and catalysts in hydrothermal vents can be captured by reduction of carbon dioxide, it is also possible that the formic acid was simply a product of the system approaching equilibrium, rather than being related to some specific property of the simulated vent minerals and putative pH gradients. Jackson (2016) performed a thermodynamic calculation to see how much formic acid would be expected to be present at equilibrium, which turned out to be ~50 micromolar. If correct, this means that the energy available in the system was not being captured, but instead the reaction was simply approaching equilibrium. Wächtershäuser (2016) also criticized the paper, noting that formaldehyde was a notorious contaminant of other published studies and that the authors had not carried out a control to rule out contamination.

Reactants → (Ce, Cy, Cn, Cd, Kn, Cl, As, En, He) time → Products

Freshwater hydrothermal fields associated with volcanic land masses are alternatives to hydrothermal vents as possible places where life began, as described in Chapter 3. These are characterized by cycles of dehydration and rehydration related to precipitation and fluctuations in water levels at the air-water-mineral interface. Figure 8.1 shows an apparatus designed to simulate such cycles in the laboratory. Results will be

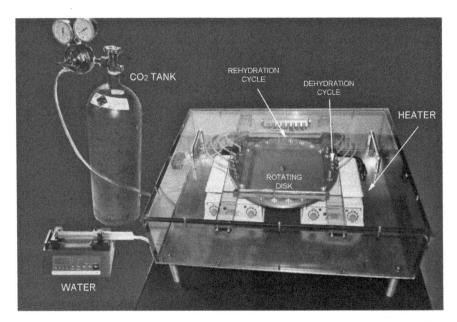

Figure 8.1. Apparatus designed to simulate a fluctuating hydrothermal field. A disk with 24 wells holds glass vials that contain reaction mixtures. The disk is heated to the desired temperature, usually 80°–90° C, and rotates so that each vial is dried for 30 minutes by a gentle stream of CO_2 gas under the eight white tubes on either side of the disk. The small syringe pump at the lower left delivers water to the vials as they pass under two tubes, one of which is indicated by an arrow. CO_2 from the tank also maintains an oxygen-free atmosphere to protect the reaction mixtures from oxidation damage.

described in detail in Chapter 9, but will be outlined here to indicate the factors that are components of the simulation.

De Guzman et al. (2014) exposed mononucleotides to multiple cycles of dehydration (Cy) in the presence of lipid as an organizing agent (En, As). The energy for the reaction (Ce) was introduced when water activity was decreased to near anhydrous conditions during drying so that ester bonds spontaneously formed and linked monomers into polymers (Cd). The reaction was also promoted by the extreme concentration of potential reactants when dried (Cn). Oligomers accumulated in a kinetic trap (Kn), and hill climbing (Cl) was introduced by cycling as monomers combined to form an increasing variety of mixed polymers.

Testing Laboratory Simulations in Natural Conditions

It is very common for researchers performing simulations of potentially prebiotic reactions to assume that the results would also be possible under prebiotic conditions. I think it is worth questioning this assumption and challenging ourselves to test it in prebiotic analogue sites. The two such sites being explored in this book are hydrothermal vents and hydrothermal fields. Unfortunately, running an experiment in an actual hydrothermal vent is a very expensive proposition because it involves diving to the vent in a specially designed submersible vehicle. On the other hand, vent proponents predict that dissolved hydrogen in alkaline vent fluid should be able to reduce CO_2 to compounds like formic acid and formaldehyde, so it is certainly possible to take samples of the vent effluent to see if such compounds are present.

Hydrothermal fields are much more accessible, and we have visited them in Kamchatka, Hawaii, Iceland, Yellowstone National Park, and Bumpass Hell on Mount Lassen in northern California. One of the simplest experiments determined whether amphiphilic compounds can assemble in hydrothermal water and encapsulate polymers (Milshteyn et al. 2018). Water samples from hot springs in Yellowstone were mixed with membrane-forming amphiphiles and RNA and put through a single cycle of evaporation and rewetting. Figure 5.7 shows the result, and it is clear that the amphiphilic mixture formed membranes and encapsulated RNA. When we repeated the experiment with seawater, the amphiphiles could not assemble into membranes. Instead they aggregated into clumps that did not encapsulate RNA. In other words, hydrothermal water from a freshwater hot spring passed this test, but salty hydrothermal seawater associated with vents did not.

Conclusions and Open Questions

Some colleagues will likely consider this analysis of the characteristics of simulations to be a bit silly. However, I will defend it by noting that it clearly illustrates why we have not yet discovered the pathway from simple mixtures of organic compounds to a living system of encapsulated molecules. None of the published simulations are sufficiently complex! The identification of the eighteen factors outlined at the beginning of this chapter is an attempt to evaluate simulations and guide future experiments in which prebiotic conditions are simulated. As illustrated by the examples presented, selected combinations of these factors are incorporated in various simulations, but cycling and encapsulation, for instance, are less commonly employed. We emphasize that cycling might be crucial to generate sufficient complexity, just as heating and cooling cycles are essential for PCR to succeed.

Regulatory feedback loops represent another factor that so far is absent from most simulations. Tsokolov (2010) argued persuasively that feedback loops are obligatory for all life today and must also have been present in the first primitive microorganisms. Engelhart et al. (2016) were among the first to test an experimental model of a feedback loop. Their approach was to encapsulate a ribozyme and oligonucleotides in fatty acid vesicles at high concentration then allow the vesicles to grow. They found that ribozyme activity was initially inhibited by the high concentration of inhibitory oligonucleotides in the vesicle but was activated by a lower concentration during vesicle growth.

> This simple physical system enables a primitive homeostatic behaviour: the maintenance of constant ribozyme activity per unit volume during protocell

volume changes. We suggest that such systems, wherein short oligonucleotides reversibly inhibit functional RNAs, could have preceded sophisticated modern RNA regulatory mechanisms, such as those involving miRNA.

References

Adamala K, Szostak J (2013) Competition between model protocells driven by an encapsulated catalyst. *Nat Chem* 5, 495–501.

Adami C, Ofria C, Collier TC (2000) Evolution of biological complexity. *Proc Nat Acad Sci USA* 97, 4463–4468.

Attwater J, Wochner A, Holliger P (2013) In-ice evolution of RNA polymerase ribozyme activity. *Nat Chem* 5, 1011–1018.

Bartel DB, Szostak JW (1993) Isolation of new ribozymes from a large pool of random sequences. *Science* 261, 1411–1418.

De Guzman V, Shenasa H, Vercoutere W, Deamer D (2014) Generation of oligonucleotides under hydrothermal conditions by non-enzymatic polymerization. *J Mol Evol* 78, 251–262.

Engelhart AE, Adamala KP, Szostak JW (2016) A simple physical mechanism enables homeostasis in primitive cells. *Nat Chem.* doi 10.1038/NCHEM.2475.

Ferris JP (2006) Montmorillonite-catalysed formation of RNA oligomers: The possible role of catalysis in the origins of life. *Phil Trans Royal Soc B* 361. doi: 10.1098/rstb.2006.1903.

Fox S (1964) Thermal polymerizations of amino acids and production of formed microparticles on lava. *Nature* 201, 336–337.

Furuuchi R, Imai E, Honda H, Hatori K, Matsuno K (2005) Evolving lipid vesicles in prebiotic hydrothermal environments. *Orig Life Evol Biosph* 35, 333–343.

Herschy B, Whicher A, Camprubi E, Watson C, Dartnell L, Ward J, Evans JRG, Lane N (2014) An origin-of-life reactor to simulate alkaline hydrothermal vents. *J Mol Evol* 79, 213–227.

Huber C, Wächtershäuser G (1997) Activated acetic acid by carbon fixation on (Fe,Ni)S under primordial conditions. *Science* 276, 245–247.

Ichihashi N, Usui I K, Kazuta Y, Sunami T, Matsuura T, Yomo T (2013) Darwinian evolution in a translation-coupled RNA replication system within a cell-like compartment. *Nat Comm* 4, 2494. doi: 10.1038/ncomms3494.

Inoue T, Orgel LE (1983) A non-enzymatic RNA polymerase model. *Science* 219, 859–862.

Jackson JB (2017) The "Origin-of-Life Reactor" and reduction of CO_2 by H_2 in inorganic precipitates. *J Mol Evol* doi: 10.1007/s00239-017-9805-9

Lincoln TA, Joyce GF (2009) Self-sustained replication of an RNA enzyme. *Science* 323, 1229–1232.

McCollom TM, Ritter G, Simoneit BRT (1999) Lipid synthesis under hydrothermal conditions by Fischer-Tropsch-type reactions. *Orig Life Evol Biosph* 29, 153–166. doi:10.1023/A:1006592502746

Miller SL (1953) A production of amino acids under possible primitive Earth conditions. *Science* 117: 528–529.

Milshteyn D, Damer B, Havig J, Deamer D (2018) Amphiphilic compounds assemble into membranous vesicles in hydrothermal hot spring water but not in seawater. *Life (Basel).* doi: 10.3390/life8020011.

Noireaux V, Libchaber A (2004) A vesicle bioreactor as a step toward an artificial cell assembly. *Proc Natl Acad Sci USA* 101, 17669–17674.

Platt, JR (1964) Strong inference. *Science* 146, 347–353.

Rodriguez-Garcia M, Surman AJ, Cooper GJT, Suárez-Marina I, Hosni Z, Lee MP, Cronin L (2015) Formation of oligopeptides in high yield under simple programmable conditions. *Nat Comm* 6, 8385.

Sousa FL, Thiergart T, Landan G, Nelson-Sathi S, Pereira IAC, Allen JF, Lane N, Martin WF (2013) Early bioenergetic evolution. *Phil Trans R Soc B* 368. doi: 10.1098/rstb.2013.0088.

Tsokolov SA (2010) Theory of circular organization and negative feedback: Defining life in a cybernetic context. *Astrobiology* 10, 1031–1042.

Wächtershäuser G (2016) In Praise of Error. *J Mol Evol* 82, 75–80.

9

Cycles, Compartments, and Polymerization

How often have I said to you that once you eliminate the impossible,
whatever remains, no matter how improbable, must be the truth.
Sherlock Holmes speaking to Dr. Watson, from
A Study in Scarlet by Arthur Conan Doyle

There is something fascinating about science. One gets such wholesale
returns of conjecture out of such a trifling investment of fact.
Mark Twain in Life on the Mississippi

Overview

The two quotes in the epigraph, in juxtaposition, always make me smile, and I tried
to keep them in mind while writing this chapter. The first eight chapters of this book
have the effect of eliminating the impossible by investigating the facts to which Twain
is referring. Perhaps he would consider them trifling, but I doubt that Twain ever
performed an experiment to test an idea. Every working scientist knows that science
is not just a set of facts but is also a set of questions. The best way to begin answering a
question is to pose a hypothesis and that hypothesis begins as a conjecture. Only when
we have a hypothesis, can we design experiments to test it, and if we are lucky, the results
of those experiments lead us a little closer to the truth.

This chapter summarizes facts that lead to an alternative scenario for life's origin in
freshwater hydrothermal conditions rather than a marine origin in saltwater hydrothermal vents. As stated in the introduction to this book, when assumptions are part
of the story they will be made explicit so that the logic that arises from them will be
clear. What follows in this overview is a list of ten prerequisites we assume are necessary
for cellular life to begin, followed by eight assumptions underlying the scenario to be
presented here.

Prerequisite conditions for life to begin:

- Dilute solutions of potential reactants are available, together with a process by which
  they can be sufficiently concentrated to react.
- Energy sources available in the environment can drive reactions such as carbon fixation, primitive metabolism, and polymerization.
- Products of reactions accumulate within the site rather than dispersing into the bulk
  phase environment.

- Amphiphiles assemble into membranous compartments over the range of temperatures, salt concentrations, and pH values related to each site.
- Biologically relevant polymers are synthesized with chain lengths sufficient to act as catalysts or incorporate genetic information.
- A plausible physical mechanism can produce encapsulated polymers as protocells then subject them to combinatorial selection.
- Organic solutes in aqueous solutions become biochemical solutes within protocells and then substrates supporting a primitive metabolism.
- Combinatorial selection leads to protocells containing polymer systems that can catalyze polymerization and specific metabolic steps.
- A cycle emerges in which one set of polymers catalyzes polymerization of a second set, and the second set guides the synthesis of the first set.
- Cellular life begins with the coevolution of primitive ribosomes, stored genetic information, and a genetic code that guides the synthesis of polymer catalysts involved in energy capture, metabolism, replication of genetic information, and division into daughter cells.

Given those prerequisites, here are the assumptions underlying the hypothetical scenario presented in this chapter. Each assumption refers to earlier chapters that described the evidence upon which the assumptions are based.

Chapters 1 and 2. Although the early Earth at the time of life's origin lacked continental land masses and was extensively covered by a salty ocean several kilometers deep, magma plumes beneath the crust generated volcanic islands resembling Hawaii and Iceland on today's Earth. An example of similar volcanism on Mars is Apollinaris Mons which was active over 3 billion years ago, in the same time frame as the origin of life on Earth.

Chapter 3. The volcanoes on early Earth were highly active with eruptions of ash and lava. However, evaporation of water from the surrounding seas fell as precipitation on the flanks of the volcanoes, then flowed downhill and accumulated in freshwater hydrothermal fields.

Chapter 4. There are multiple potential sources of biologically relevant organic compounds. Three possibilities include extraterrestrial infall in the form of meteoritic dust and impacting comets; atmospheric synthesis of compounds like formaldehyde and HCN that react to produce amino acids, nucleobases, and sugars; and geochemical synthesis of similar organic compounds as volcanic gases pass through hot minerals. Because oxygen is absent, the organic compounds accumulate over much longer time spans than they would in today's oxidizing atmosphere.

Whatever the source, if the organic compounds happened to fall or be dispersed into the ocean they would dissolve to form a very dilute solution. In contrast, those same compounds falling on or synthesized in volcanic land masses exposed to the atmosphere would become concentrated on the mineral surfaces rather than dispersed. Precipitation would flush the compounds into pools of hydrothermal water.

The circulating water in hydrothermal fields was moderately acidic, with pH ranging from 2 to 5. The concentrations of common cations such as sodium, potassium, calcium, and magnesium would be in the millimolar range, with accompanying anions of silicate, sulfate, and chloride. The water would not be stagnant but would instead circulate and undergo cycles of evaporation to a dry state in response to heat and low relative humidity followed by rehydration related to precipitation, hot springs, and geyser activity. The cycling frequency could be as short as minutes when rehydration was caused by geysers splashing on surrounding hot mineral beds; could last hours due to fluctuating levels of pools fed by hot springs; or could stretch to days or weeks when related to cycles of complete evaporation followed by precipitation.

Chapter 5. During the dry phase of a cycle, films of concentrated organic compounds were deposited on mineral surfaces where they could undergo reactions required for the origins of life. If amphiphilic compounds were present in the mixture, self-assembly of organized multilamellar structures would occur in the concentrated films on mineral surfaces. This means that the solutes would not simply go from a fluid state dissolved in water to a solid state in the films. Instead, the films are liquid crystals in which the solutes dissolve and diffuse as highly concentrated and organized reactants within the two-dimensional planes of multilamellar structures.

Chapter 6. A variety of polymers would be synthesized from monomers by condensation reactions in the dry phase, driven by low water activity and linked by ester and peptide bonds. The reactions occur between bilayers of multilamellar structures formed by amphiphilic compounds present in the mixture and within eutectic phases produced from solutes during the evaporation process.

Chapter 7. Upon rehydration, the polymers are encapsulated in microscopic membranous vesicles. Each vesicle is different from all others and therefore can be considered to serve as an experiment in a natural version of combinatorial chemistry. If certain combinations of organic solutes are captured in the vesicles, they will be sufficiently concentrated to undergo chemical reactions related to primitive metabolic pathways. They will also contain polymer species that are produced during the dry phase of the cycle, a few of which happen to catalyze reactions that capture chemical energy and transform potential nutrients into compounds supporting protocellular growth. Those protocells will tend to persist during cycling while the components of inert protocells are recycled.

Chapter 8. The processes described here can be simulated in the laboratory. Various simulations have been developed, each related to a specific step in the pathway to the origin of life.

This chapter. As a first step toward demonstrating feasibility of the pathway proposed here, it is essential to show how such polymers could be synthesized by nonenzymatic reactions and then encapsulated to produce protocells. This chapter describes progress toward that goal.

Can Cycles of Dehydration and Rehydration Drive Protocell Assembly?

A valid hypothesis must be subjected to critical experimental tests, so the focus here is to describe the tests already performed, and the results.

Test 1.

Hypothesis: Amphiphilic compounds can be dispersed in aqueous phases as membranous vesicles. If a mixture of vesicles and monomers is exposed to cycles of dehydration and rehydration, the vesicles fuse and trap the monomers in multilamellar matrices.

The idea that polymerization can be driven simply by drying and heating is obvious and was one of the first to be submitted to experimental testing (Fox and Harada, 1958; Lohrmann and Orgel, 1971; Lahav et al., 1978). But, even though polymerization can occur under these conditions, the polymeric material generated from hot, dry, and melted amino acids was a complex mass of cross-linked and degraded products organic chemists call tar (Benner et al., 2012) that had no clear way to evolve into catalysts or undergo replication. Furthermore, there is a general distaste among chemists for reactions requiring a dry state. Solution chemistry is much preferred, and in fact, living cells can be described as microscopic compartments in which reactions occur in aqueous solutions.

The scenario we are proposing does not incorporate a dry state in which the potential reactants exist as a solid. Instead, if amphiphilic compounds are present together with monomers, evaporation causes both the monomers and the vesicles to become so concentrated that they fuse into a multilamellar matrix in which the monomers are concentrated between two-dimensional planes of bilayers. Because the amphiphilic matrix is a liquid crystal, an aqueous solvent has been replaced by a dehydrated liquid crystalline solvent in which reactants are free to diffuse and undergo polymerization reactions (Plate 9.1).

This process has been experimentally confirmed by X-ray diffraction (Toppozini et al., 2013; Himbert et al., 2016) and neutron scattering analysis (Misuraca et al., 2017) of multilamellar films that are produced when aqueous mixtures of a mononucleotide and phospholipid vesicles undergo evaporation and fusion.

Test 2.

Hypothesis: Because the monomers are extremely concentrated within the matrix and water activity is reduced, condensation reactions drive polymerization and the resulting polymers can be visualized by standard methods of gel electrophoresis. Furthermore, if the polymers resemble RNA, they should be recognized by enzymes that are specific for biological RNA.

The initial tests were performed by Rajamani et al. (2008) in a relatively simple apparatus consisting of test tubes in a laboratory heating block. Each test tube was sealed with a stopper in which a tube was inserted so that a gentle stream of CO_2 could flow into a small volume of AMP—in its acid form rather than a sodium salt. The effect of having several different lipids—including phosphatidylcholine, phosphatidic acid, and lysophosphatidylcholine—in the mix was tested. The mixture was dried for two hours, rehydrated by hand-injecting water through the stopper, then dried again, with the cycle repeated up to seven times. The synthesis of polymers was confirmed with the fluorescent dye RiboGreen that was developed for RNA analysis. The products were isolated by a standard method in which ethanol causes RNA to precipitate into a pellet during centrifugation.

The RiboGreen assay was encouraging because it showed that a polymer had been synthesized which behaved like RNA. For the next test, the products were treated by a standard method to label RNA with radioactive phosphate. This involved two enzymes, alkaline phosphatase to remove phosphate groups from the ends of RNA-like molecules, followed by T4 kinase which can transfer radioactive phosphate from ATP to the ends. The labeled products were separated by gel electrophoresis and visualized by a device that produces an image of the gel and any radioactive products.

Several gels are shown in Figure 9.1; each lane reveals products synthesized under different conditions in individual tubes. It is obvious that they were labeled with radioactive phosphate and show up as dark bands in each gel. The fact that they have been labeled indicates that the two enzymes recognized the end groups as RNA, which increased our confidence that an RNA-like product was synthesized.

There are several lessons to be taken from this initial series of experiments. First, the yields increased with each cycle (Panel A) and the products ranged in length from 10 to over 100 nucleotides in length when compared with the ladder on the right of panel A. Elevated temperature was required for the reaction (Panel B) and all three lipids promoted the reaction (Panel C). The result for lysophosphatidylcholine (LPC) was particularly interesting because there was an indication that single nucleotide additions were resolved. For this reason, LPC was chosen for experiments to be described later.

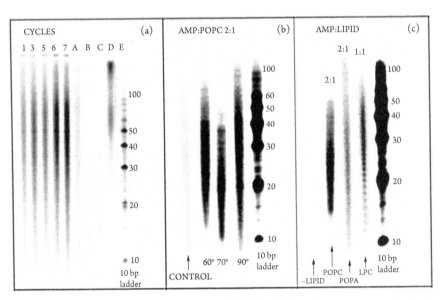

Figure 9.1. Gel electrophoresis analysis reveals that RNA-like polymers were synthesized from mononucleotides by wet–dry cycling. The products were labeled with radioactive phosphate and show up as dark bands. Panel A. The amount of product increases with the number of cycles (from 1 to 7), with products ranging from 10 to over 100 nucleotides in length. The lanes marked A, B, and C are several different negative controls. Lane D is a positive control performed with polyadenylic acid, a synthetic homopolymer of RNA. Lane E is a ladder produced by RNAs of known lengths from 10 to 100 nucleotides. Panel B. The control was an unheated reaction mixture dried at room temperature. The other lanes show products heated and cycled at 60, 70, and 90° C. Panel C. Three different lipids that promote polymerization are shown. Note the ladder in the lane marked LPC (lysophosphatidylcholine) in which products are resolved with single nucleotide increments in length. Adapted from Rajamani et al. 2008.

Rajamani et al. were also the first to use nanopore analysis to test whether linear anionic polymers were present among the products. The results were positive, and nanopore analysis was used in later experiments to confirm that single-stranded RNA-like products were present in the mixture of products. The nanopore method and results will be described in detail in the next section.

Confirming the Nature of the Polymers

Because of the success of the initial experiments, over the next few years the automated chamber described in Chapter 8 was fabricated to simulate the hydration–dehydration cycles associated with geothermal hot springs and fluctuating pools. There were two differences from the original experiment, one of which was that AMP was mixed with UMP, its base-pairing partner in RNA. There would never have been a pure nucleotide in the prebiotic environment, so this was one step toward greater complexity. The second reason is more interesting. If both AMP and UMP are present and become polymerized, there is a chance that base pairing could occur, most likely as hairpins in single strands but also possibly as duplex regions forming between two strands. We also predicted that strands with duplex portions would strongly interact with intercalating dyes during gel electrophoresis so that we could avoid expensive and time-consuming radioactive labeling.

Test 3.

Hypothesis: The polymers that were labeled by radioactive phosphate should also be stained in gels by intercalating dyes such as ethidium bromide.

The results that follow were reported by DeGuzman et al. (2014) and came from experiments carried out in the simulation chamber described in Chapter 8. In a typical experiment, glass vials containing reaction mixtures to be tested were placed in the wells around the perimeter of an aluminum disk. The chamber was filled with CO_2, an inert gas, to produce and maintain anaerobic conditions; the disk was heated to a desired temperature, usually 85° C; and rotation of the disk was controlled by a programmed stepper motor. As the disk rotated, each sample was exposed to dehydration under a flow of dry CO_2, followed by rehydration when the rotation brought a vial under each of two ports delivering water from a syringe pump. In a typical experiment, the rate of rotation caused each sample to undergo one hydration–dehydration cycle every 90 minutes. We chose 2-hydroxy LPC as an organizing matrix: it forms micelles in dilute solutions but assembles into multilamellar structures upon drying, thereby avoiding the necessity for preparing liposomes as an intermediate step and making it convenient to use.

A typical reaction mixture in the vials undergoing wet–dry cycles in the simulation chamber had 0.2 ml of 10 mM mononucleotides and 0.1 ml of 10 mM LPC. The polymer products were isolated either by precipitation in ethanol followed by centrifugation or with spin tubes developed for isolating RNA oligomers. The products readily formed pellets or were adsorbed, a property consistent with polymers that behave like RNA.

Agarose gels were used to visualize polymeric products. The gels contained ethidium bromide dye which displays enhanced fluorescence upon intercalation between stacked bases of nucleic acid species. Figure 9.2A shows a typical 4% agarose gel with a ladder of RNA having known sizes to indicate polymer lengths in terms of nucleotide number. (The image is inverted to show fluorescence more clearly as dark against a light background.) Note that the dye does not stain single-stranded polyadenylic acid (polyA) or polyuridylic acid (polyU) because in the absence of duplex strands with stacked bases, it is unable to intercalate efficiently. However, the dye intensely stains the duplex strands that are produced when polyA and polyU are mixed in 1:1 ratios to form duplex species.

Significantly, the RNA-like product was also stained, indicating that a certain amount of duplex structure was present, probably as intramolecular hairpins. Note also that the stained polymer is in the same range of lengths as the polymers labeled with radioactive phosphate. To check whether the labeling was an artifact of ethidium bromide staining, the products were also analyzed using a different dye. Figure 9.2B shows a 2% agarose gel with the SYBR-Safe dye which is less dependent on the presence of duplex strands. The product moves further in the lower concentration gel but is again stained.

Test 4.

Hypothesis: If the products are polymers of mononucleotides they will be linear polyanions that produce ionic current blockades when analyzed by a nanopore technique.

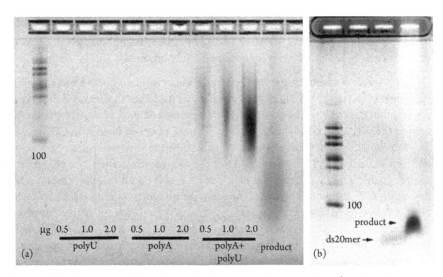

Figure 9.2. A. Precast gel showing synthetic homopolymers of RNA (polyU, polyA, and polyA+polyU duplex). The images are inverted so that fluorescence shows up as dark against a light background. Only the duplex polymer is stained by ethidium bromide, an intercalating dye that is incorporated in the gel. The first lane on the left is an RNA ladder with the 100mer marked, and the last lane on the right shows that the product from an experimental HD reaction of AMP + UMP + lipid is also stained by the ethidium dye. B. The experiment was repeated with a 2% agarose precast gel containing the SYBR-Safe stain. A double stranded 20mer of RNA was stained, as expected, and so was the longer product of the polymerization reaction. Adapted from De Guzman et al. 2014.

Alpha hemolysin is a 22,000 dalton protein antibiotic secreted by *Staphylococcus aureus*. When added to a solution bathing a lipid bilayer, the protein subunits partition into the membrane and spontaneously assemble into a heptameric transmembrane channel with a diameter of 2 nm. When a voltage of 100 mV or more is applied across the bilayer, negatively charged nucleic acid molecules in solution are captured by the electric field in the channel and drawn through in a process called single molecule electrophoresis (Kasianowicz et al., 1996; Deamer et al., 2016). The translocation is very fast, in the range of 2–20 microseconds per nucleotide in the polymer. During translocation, individual nucleic acid strands fill the nanopore and produce a characteristic ~90% ionic current blockade (Fig. 9.3).

Some results from nanopore analysis are shown in Figure 9.4. Abundant ionic current blockades were produced by the polymer products synthesized from a range of mononucleotides, including AMP or UMP alone as well as mixtures of AMP and UMP. Only linear polyanions such as RNA and DNA are known to produce ionic current blockades in the hemolysin channel, so the nanopore results confirm that wet–dry cycles can drive condensation reactions leading to polymerization of linear RNA-like polymers.

Test 5.

Hypothesis: If the polymers are linked by phosphoester bonds, they should be hydrolyzed by RNAse or exposure to alkaline pH ranges.

Da Silva et al. (2014) found that addition of monovalent salts such as NaCl, KCl, and NH_4Cl promoted polymerization of mononucleotides even in the absence of lipids. The polymers of AMP and UMP were indistinguishable in their properties from those synthesized in lipid matrices and provided important clues to the nature of the linking bonds. For instance, the polymers could be partially hydrolyzed by RNAse, which meant that a fraction of the linking bonds were the 3'–5' bonds of biological RNA. This would be expected because the ribose of RNA has hydroxyl groups on its 2' and 3' carbons and both would react with the phosphate of a second nucleotide.

Test 6.

Hypothesis: All double-stranded nucleic acids "melt" at characteristic temperatures when heated. If the monomers used to synthesize the polymers are capable of forming Watson-Crick base pairs stabilized by hydrogen bonds, they should exhibit hyperchromicity when exposed to increasing temperature.

When duplex nucleic acids are heated, they undergo a process referred to as melting in which the strands become separated. When this happens, a characteristic increase in the absorbance of ultraviolet light at 260 nm wavelength occurs. If hyperchromicity is observed in the products, it would indicate that not only were polymers being synthesized, but the strands had duplex portions in which AMP was hydrogen bonded to UMP. Figure 9.5A shows hyperchromicity of a known duplex RNA structure, a 1:1 mixture of polyA and polyU. Hyperchromicity is obvious over the temperature range of 20 to 60° C. The reason the range is so broad is that the polyA and polyU have different chain lengths so that the duplex species also vary in length. The hyperchromicity that begins at 20° C is related to shorter-chain duplex structures and the end point of 60° C reflects longer duplexes. Figure 9.5B shows hyperchromicity clearly evident in

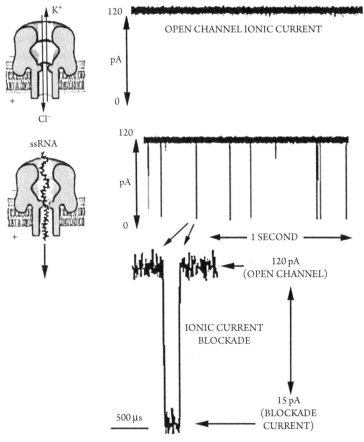

Figure 9.3. The basic principle of nanopore analysis is that nucleic acid strands can be drawn by an applied voltage through nanoscopic pores such as hemolysin. During translocation, the strand blocks ~90% of the current of potassium and chloride ions in the solution of 1.0 M KCl that bathes the nanopore.

a polymer produced by cycling a mixture of AMP and UMP but absent in a polymer synthesized from AMP alone. These results are consistent with the presence of duplex structures in the polymer synthesized from AMP and UMP, probably as hairpins in single strands of the product.

RNA-Like Polymers Are Not Biological RNA

From these results, we concluded that in the presence of an organizing matrix, wet–dry cycles can drive polymerization of mononucleotide mixtures, resulting in polymers ranging from 10 to >100 nucleotides in length. However, the polymers are not equivalent to biological RNA which is synthesized from nucleoside triphosphates, a process catalyzed by an RNA polymerase that uses a DNA template to specify the

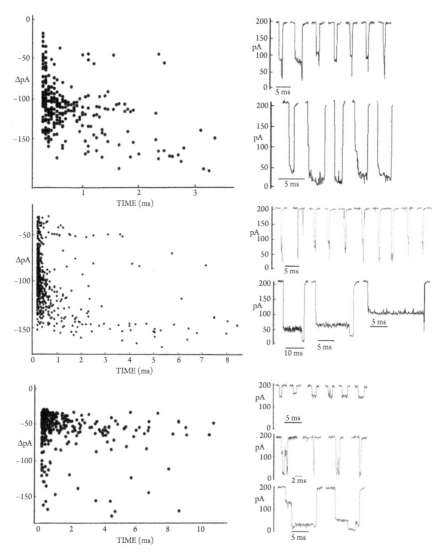

Figure 9.4. Event diagrams and blockade patterns for control 20mers of ssRNA and dsRNA molecules (A and B) and for polymeric products synthesized by exposing a mixture of AMP, UMP, and LPC to multiple wet–dry cycles (C). Each dot represents a single molecule traversing the nanopore, with examples of individual blockades of shorter and longer events shown on the right. Note that the amplitude scale is the same, while duration scales are different for each type of molecule examined. The event diagram of ssRNA 20mers shows amplitude and duration of ionic current blockades when single strands are captured by the nanopore. The event diagram for double-stranded RNA 20mer also reveals short and long blockades, with examples of individual events shown on the right. The event diagram for products of AMP/UMP/lipid mixture reveals both short and long duration events that indicate the presence of single-stranded polymers, some of which have intramolecular hairpins that slow translocation until they unzip. Adapted from De Guzman et al. 2014.

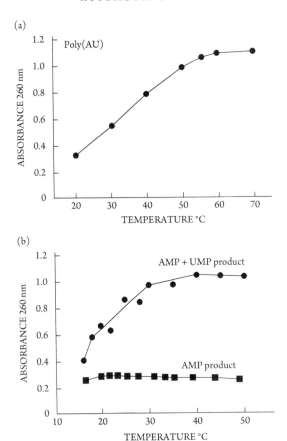

Figure 9.5. Hyperchromicity of RNA and RNA-like polymers. A. The UV absorbance (260 nm) of an RNA double helix increases with temperature when the duplex structure melts into single strands. In this case, the RNA duplex was formed by mixing synthetic homopolymers called polyA and polyU in a 1:1 ratio. This mixture is known to assemble into a double helical structure. B. The polymer produced from a mixture of AMP and UMP also exhibited hyperchromicity with increasing temperature, but a polymer synthesized from AMP alone under the same conditions did not. The duplex structure of the AMP + UMP polymer was probably in the form of hairpins within the chain rather than the double helix of the poly(AU) shown in A. Adapted from De Guzman et al. 2014.

base sequences. In biological RNA, the nucleotide monomers are linked entirely by phosphoester bonds between nucleotides referred to as 3'–5' bonds because the phosphate links the 3' carbon of one ribose to the 5' carbon of the next. Because ribose also has a hydroxyl group on its 2'-carbon, the nonenzymatic synthesis has an equal chance of producing a 2'–5' bond, which is the reason we refer to the polymer as RNA-like. Another difference is that biological RNA has specific chemical groups at the 3' and 5' ends of the molecule, while the end groups are unspecified in the RNA-like products. We also found that some of the bonds cannot be hydrolyzed by RNAse or alkaline hydrolysis, which means that bonds other than ester linkages are present in the polymers.

Test 7.

Hypothesis: If a deoxyribonucleotide is used as a monomer, some of the products will be equivalent to biological DNA.

In order to test whether deoxyribonucleotides can also undergo polymerization, we exposed 5'-thymidine monophosphate (TMP) to multiple wet–dry cycles. Because deoxyribonucleotides can only be linked by 3'–5' bonds, our expectation is that in the trillions of molecules synthesized there should be some that can be recognized by enzymes for which DNA is a substrate.

TMP (10 mM) was mixed with an organizing agent and put through four hydration–dehydration cycles under the conditions described earlier in this chapter for ribonucleotides. Two kinds of organizing matrix were used: LPC, which produces a multilamellar liquid crystalline matrix when dried; and ammonium chloride, which crystallizes during evaporation. Because other solutes are excluded by crystallization, the mononucleotides form a highly concentrated eutectic phase between crystals in which condensation reactions can occur (Da Silva et al., 2014).

The reactions were performed by exposing 0.1 mL samples to multiple cycles of hydration–dehydration at 30-minute intervals (DeGuzman et al., 2014). After four such cycles at 85° C, the polymerization products were isolated by ethanol precipitation and labeled with radioactive phosphate using the method described by Rajamani et al. (2008).

Figure 9.6 shows a gel of the radioactively labeled products that used ammonium chloride as an organizing agent along with a scan of the gel. It is clear that labeled bands ranging from 32 to 90mers were present, the same range that was seen with the ribonucleotide monomers in earlier work.

If DNA has actually been synthesized as polymers of TMP, it should be possible to detect the polymers by nanopore analysis with the MinION sequencing device. Therefore, we used enzymes called ligases to attach the polymers to a known DNA strand. The blockade signal from the MinION was expected to show the known DNA strand together with a strand that was composed of polymerized TMP, referred to as oligoT homopolymer. When the blockades began to appear on the computer screen, many of them looked like the example shown in Figure 9.7:

Base calling was performed on signals having the apparent homopolymer signature. One example is shown here with the homopolymeric portion indicated in bold font followed by the sequence of the synthetic DNA ligated to it:

TTTTTTTTTTTTTTTTTTTTTTTTTTTTTTTTTTTTTTTTTTTTTTTTTTTTACTTCGTTCAGTT
ACATGTGCTGAGTTCAATCAAGGTTCAATCAAAGGTTCAATCAAAGGTT
CAATCAGAGTTTCAATCAAAGGTTCAATCAAGGTTCCATCAAGGTTCA
ATCAATCAAAGAAGGGTTCAATCAAGGATTCAACGGGGTTTGGTAGGG
TTTCCATTCGGAGAATTTTGGTCAGGCTGCATTGGCTGCTTGGTGAGGT
TTGTCGGGGTTCCGGTCAGGGTTTCCGTCGAGAGTTCCGGTCAGGGT
TTTGTCGAGGTTCCGATGAAA

For us, this is convincing evidence that a process as simple as wet–dry cycling can synthesize nucleic acids. Particularly convincing is the fact that the polymers are recognized not just by the enzymes involved in 32-P labeling but also by three

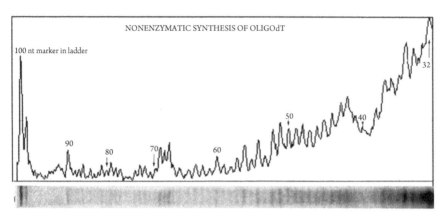

Figure 9.6. OligoT (oligothymidylic acid) was synthesized by four wet–dry cycles, followed by labeling with radioactive phosphorus and separation of labeled products by gel electrophoresis. The gel lane with dark bands shows the labeled oligoT with a 100 nucleotide marker at the far left. The gel image was scanned to reveal the approximate chain lengths of the bands, which are separated by single nucleotide resolution.

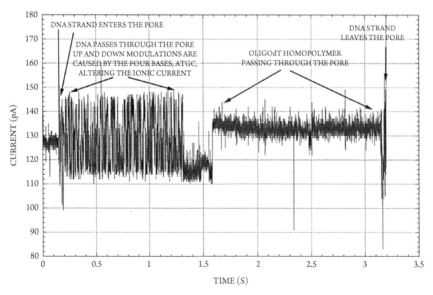

Figure 9.7. This example shows an electrical signal produced when a single molecule of DNA ligated to the oligoT polymer is drawn through a nanopore in the MinION sequencing device. The signals can be decoded by a base calling algorithm; one example is described in the text.

enzymes that catalyze reactions required for preparing DNA for MinION sequencing, including the ligase and the helicase motor that advances the strand through the nanopore.

Test 8.

Hypothesis: If the polymers are synthesized in the presence of a membrane-forming lipid, they should accumulate in vesicles upon rehydration and be visualized by dyes that bind to nucleic acids.

This test was performed using dAMP and TMP, two nucleotides of DNA that can only form 3'–5' bonds. We also used decanoic acid and its monoglyceride rather than a phospholipid in order to conduct the experiment with a simpler amphiphilic mixture that would be more plausible in prebiotic conditions. A fluorescent dye called DAPI was added because it is commonly used to label DNA in cells. We reasoned that if a polymer of dAMP and TMP was synthesized and accumulated in vesicles, the dye would reveal the polymer by its strong fluorescence. Plate 9.2 shows the vesicles by phase and fluorescence microscopy after 1, 2, 3, and 4 cycles. The vesicles clearly contain a fluorescently stained polymer that increased with each cycle.

Can Random Polymers Evolve Catalytic Functions When Exposed to Cycles of Selection and Amplification?

The prebiotic synthesis of nucleic acids must have been nonenzymatic and would lead to complex mixtures of variable-length strands having random sequences of nucleotides. How can they possibly begin to incorporate catalytic functions? This fundamental question can be addressed by recalling results published by David Bartel and Jack Szostak (1993). Their aim was to establish an experimental system in which random polymers of RNA were subjected to cycles of selection and amplification to test the possibility that catalytic polymers might emerge. The catalytic property chosen was ligase activity in which an RNA ribozyme catalyzes its linkage to another RNA strand that is attached to small beads in a column. To perform the experiment, RNA strands in the range of 300 nucleotides in length were first synthesized by a reaction that incorporated the four nucleotide monomers in random sequences. The result was a few micrograms of RNA containing trillions of molecules, each with a base sequence different from all the others. The linear strands would tend to fold into three dimensional structures, a rare few of which might happen to have weak ligase activity. If so, these would be expected to adhere to the column because they formed phosphoester bonds with the RNA that was bound to the beads in the column. Although the ester bonds would all be equivalent, some of the polymers would more efficiently catalyze the reaction and therefore would bind more extensively to the beads.

In the actual experiment, the products were cycled through the process multiple times and were visualized by gel electrophoresis. At the start of the experiment, only the 300mer could be seen at the top of the gel. In the fourth and fifth rounds, shorter bands began to appear because certain catalytic species were being selected and amplified. With further cycling, some of those species began to disappear while others become prominent. After 10 cycles, three new bands were visible that are the "winners" of the competition, and the ligase activity had increased over 7 million-fold!

To summarize: at first, every molecule of RNA was approximately the same length but different in sequence from all the rest, but then a selective hurdle was imposed in the form of a ligation reaction that allowed only certain molecules to be selected, then released and amplified. Over 10 cycles of selection and amplification, specific catalytic molecules emerged by a process reflecting selection and evolution at the molecular level. The inescapable conclusion is that sequence-specific information and catalytic functions can appear in mixtures containing large numbers of random-sequence RNA from which specific sequences having a catalytic property can be selected and amplified.

But this was a laboratory experiment using purified nucleoside triphosphates and reactions catalyzed by multiple enzymes. Is it possible that similar selective processes could have occurred on the prebiotic Earth when the first self-assembled systems of encapsulated random-sequence polymers assembled by chance? That question is at the edge of our current knowledge, and further experiments will be required before we know the answer. In Chapter 10 we will describe a process by which selection and evolution of encapsulated polymers could occur and also some experimental tests of the hypothesis.

References

Bartel DB, Szostak JW (1993) Isolation of new ribozymes from a large pool of random sequences. *Science* 261, 1411–1418.

Benner SA, Kim H-J, Carrigan MA (2012) Asphalt, water, and the prebiotic synthesis of ribose, ribonucleosides, and RNA. *Acc Chem Res* 45, 2025–2034.

Da Silva L, Maurel MC, Deamer D (2014) Salt-promoted synthesis of RNA-like molecules in simulated hydrothermal conditions. *J Mol Evol* 80, 86–97.

Deamer D, Akeson M, Branton D (2016) Three decades of nanopore sequencing. *Nat Biotech* 34, 518–524.

De Guzman V, Shenasa H, Vercoutere W, Deamer D (2014) Generation of oligonucleotides under hydrothermal conditions by non-enzymatic polymerization. *J Mol Evol* 78, 251–262

Fox SW and Harada K (1958) Thermal copolymerization of amino acids to a product resembling protein. *Science* 128, 1214.

Himbert S, Chapman M, Deamer DW, Rheinstadter MC (2016) Organization of nucleotides in different environments and the formation of pre-polymers. *Sci Rep* 6. doi:10.1038/srep31285.

Kasianowicz J, Brandin E, Branton D, Deamer DW (1996) Characterization of individual polynucleotide molecules using a membrane channel. *Proc Natl Acad Sci USA* 93, 13770–13773.

Lahav N, White D, Chang S (1978) Peptide formation in the prebiotic era: Thermal condensation of glycine in fluctuating clay environments. *Science* 201, 67–69.

Lohrmann R, Orgel LE (1971) Urea-inorganic phosphate mixtures as prebiotic phosphorylating agents. *Science* 171, 490–494.

Misuraca L, Natali F, da Silva L, Peters J, Demé B, Ollivier J, Seydel T, Laux-Lesourd V, Haertlein M, Zaccai G, Deamer D, Maurel MC (2017) Mobility of a mononucleotide within a lipid matrix: A neutron scattering study. *Life (Basel)*. doi: 10.3390/life7010002.

Rajamani S, Vlassov A, Benner S, Coombs A, Olasagasti F, Deamer D (2008) Lipid-assisted synthesis of RNA-like polymers from mononucleotides. *Orig Life Evol Biosph* 38, 57–74.

Toppozini L, Dies H, Deamer DW, Rheinstädter MC (2013) Adenosine monophosphate forms ordered arrays in multilamellar lipid latrices: Insights into assembly of nucleic acid for primitive life. *PLoS ONE* 8, e62810.

10

Integrating Chemistry, Geology, and Life's Origin

COAUTHORED WITH BRUCE DAMER

In so far as a scientific statement speaks about reality, it must be falsifiable; and in so far as it is not falsifiable, it does not speak about reality.

Karl Popper

Overview

Chapter 8 recalled John Platt's recommendation that testing alternative hypotheses is a preferred way to perform research rather than focusing on a single hypothesis. Karl Popper proposed an additional way to evaluate research approaches, which is that a strong hypothesis is one that can be falsified by one or more crucial experiments.

This chapter proposes that life can begin with chance ensembles of encapsulated polymers, some of which happen to store genetic information in the linear sequences of their monomers while others catalyze polymerization reactions. These interact in cycles in which genetic polymers guide the synthesis of catalytic polymers, which in turn catalyze the synthesis of the genetic polymers. At first, the cycle occurs in the absence of metabolism, driven solely by the existing chemical energy available in the environment. At a later stage, other polymers incorporated in the encapsulated systems begin to function as catalysts of primitive metabolic reactions described in Chapter 7. The emergence of protocells with metabolic processes that support polymerization of self-reproducing systems of interacting catalytic and genetic polymers marks the final step in the origin of life.

The above scenario can be turned into a hypothesis if it can be experimentally tested—or falsified, as described in the epigraph. The goal of falsification tends to be uncomfortable for active researchers. It's a very human tendency to be delighted with a creative new idea and want to prove it correct. This can be such a strong emotion that some fall in love with their idea and actually hesitate to test it. They begin to dislike colleagues who are critical and skeptical. However, my experience after 50 years of active research is that we need to think of our ideas as mental maps and expect that most of them will not match the real world very well. And so, I say to my students, "When you have a new idea it's OK to enjoy it and share it with others, but then you must come up with an experiment that lets you discard it. The rare ideas that survive critical experimental testing are the only gems worth keeping, and you should find them as soon as possible by discarding all the ideas that fail testing."

The practice of falsification works best in physics because a single prediction and its experimental test can often be performed. An early example of alternative hypotheses in cosmology was Fred Hoyle's suggestion that the universe exists indefinitely in a steady state, while Robert Dicke and George Gamov independently proposed in 1946 that if the universe began as a Big Bang there should be a measurable cosmic background radiation. The detection of the predicted radiation by Arno Penzias and Robert Wilson in 1964 led to a consensus that the universe did have a beginning, now estimated to be 13.7 Gya. It was a revelation to realize how ancient our planet actually is, one third the age of the universe itself.

Life is such a complex phenomenon that falsification of theories in areas of study related to it is not as simple as it can be in physics. Nonetheless, rather than devising one crucial experiment, we can perform multiple experimental tests and use the weight of accumulating evidence to decide plausibility and explanatory power. Proposals for an origin of life in salty ocean water at hydrothermal vents or in freshwater at hydrothermal fields can be considered to be alternative hypotheses. Our research has focused on testing a freshwater origin, and in this chapter, we will summarize previous attempts to falsify the hypothesis then make predictions that can be tested experimentally to guide future research efforts. Our goal can be summarized in a single sentence: *Confirm or falsify a nonenzymatic process by which catalytic and replicating systems of polymers are produced under plausible prebiotic conditions that support cycles of combinatorial selection in which encapsulated systems of polymers emerge and begin to evolve.*

We will begin this chapter with a narrative that integrates the chemistry of Chapter 9 with what we know about the geology of the early Earth and its ocean. The narrative is clearly classified as a conjecture, as it represents a construct of ideas based on facts, rather than the facts about chemistry presented in Chapter 9. We will then go on to describe some predictions and experimental tests that are designed to turn conjecture into a working hypothesis about how protocells can take the next step toward the origin of life.

Encapsulated Systems of Random Polymers Can Potentially Undergo Combinatorial Selection

Exposing encapsulated systems of random polymers to selection in a natural environment is very different from the laboratory experiments described in Chapter 9 because the selective process must work not just on populations of molecules in solution but also on the compartments in which they are encapsulated. Furthermore, the polymers are not being selected for a single function but for multiple functions that are properties of encapsulated systems of polymers. Although this is obviously complicated, it is possible to make a partial list of such selective factors and then imagine how encapsulated polymers can respond to the imposed stresses.

What follows is a list of stresses that represent selective hurdles to be overcome by populations of protocells and also some possible responses to those stresses.

Physical variables

- Temperature
- Wetting and drying
- Osmotic pressure
- Shear forces
- Dispersal and dilution
- Influence of dissolved or suspended minerals

Chemical variables

- Decomposition rates
- Energy sources
- Nutrient concentrations
- pH and dissolved ion concentrations
- UV light and radiation

Possible ways that protocells can evolve by responding to environmental stresses

- Increased structural stability of protocell compartments in which polymers interact with bilayer membranes
- Varying the composition of amphiphilic compounds
- Insertion of polymeric pores to reduce osmotic effects and allow uptake of potential nutrients
- Polymers that can catalyze primitive metabolic reactions
- Polymers that have simple polymerase activity such as template directed replication
- Polymers that can use energy and nutrients to grow and reproduce

The list of potential stresses makes it obvious that protocell selection in a natural environment is immensely complex. We will not attempt to discuss all of the stresses but instead will focus on some that are relevant to the hypothesis we are developing—those related to the physical stability of membranous compartments containing encapsulated polymers—and which are also relatively accessible to experimental testing.

Stresses During Wet–Dry Cycles
Are Selective Hurdles

Several physical stresses develop in membranous compartments and their contents as the aqueous medium evaporates. The most obvious stress is that ionic solutes become increasingly concentrated, as do any organic compounds that may be present. For instance, typical hydrothermal fields have monovalent ions such as sodium and potassium chloride present in approximate 10-15 mM concentrations (see Table 3.1).

Now imagine that evaporation removes 99% of the water. It is clear that the solutes become much more concentrated, exceeding 1.0 M salts. Can membranous compartments and their contents withstand such extreme conditions? The internal concentrations also increase dramatically during drying and when rehydration occurs, for instance during precipitation, a huge osmotic gradient is transiently produced across the membranes. Unless the osmotic pressure can be compensated for in some way, vesicles could swell and perhaps burst, releasing their contents to the surrounding bulk medium. If this actually happens, there is no way to preserve the systems of polymers the hypothesis requires.

We have begun to investigate possible compensating factors and found that lipid vesicles do not actually burst. Instead, upon rehydration the membranes are initially present as multilamellar structures, shown in Plate 10.1. The outer layers are first to experience an osmotic gradient, and they do swell as water penetrates the membrane. But instead of bursting, surrounding lipids move into the growing membranes which begin to bud off as vesicles. At some point the internal ionic concentration balances the external concentration, and the vesicles are stabilized. There is some loss of ionic solutes across the membranes through transient defects in the membranes, but the defects are not large enough to accommodate longer polymers which remain encapsulated in the budded off vesicles. The inset in Plate 10.1 shows a lipid film that was dried in the presence of short strands of DNA (~600mers); it is clear that the DNA polymer has remained in the resulting vesicles.

Significantly, in the next cycle of evaporation the vesicles will aggregate and finally fuse, a process in which concentrated solutes are captured between lipid bilayers and mixed again with the polymers. In this way, cycling provides a continuous series of opportunities for solutes and polymers to interact and undergo further reactions that increase the complexity of the contents. This is an important difference to reactions that are proposed to occur in solution in noncycling conditions, which tend to approach an equilibrium in which no further synthesis occurs.

Even though the vesicles and their cargo of polymers may survive the stresses associated with cycling, they are still in an environment that has multiple other stresses, so we propose that the stabilization of lipid bilayers by functional polymers is likely to be a primary selective factor. This is our first test, because in the absence of such polymers the lipid membranes and encapsulated systems of polymers are much too fragile to persist over multiple cycles.

Evolution of Protocell Populations: Coupled Phases

Plate 10.2 illustrates how we expect protocells to respond to a cycle of dehydration followed by rehydration in a natural environment. Keep in mind that a variety of polymers are being synthesized from monomers like amino acids and nucleotides, and that the system operates in an indefinite series of cycles that pump the components away from equilibrium toward a steady state in which the polymers are maintained in a kinetic trap. The cycling drives serial, natural experiments that undergo combinatorial selection in the form of encapsulated polymers. We refer to this process as "coupled phases" (Damer and Deamer, 2015) because during each cycle, polymers are exposed to three distinct phases that involve self-assembly properties of amphiphilic compounds: a multilamellar phase characterized by reduced water activity; a hydrated phase in which encapsulated polymer mixtures are released into an aqueous bulk within vesicles; and an intermediate gel-like phase in which surviving protocells and concentrated solutes aggregate before fusing back into a dehydrated multilamellar phase.

Most of the protocells in a given cycle will be inert and soon disrupted by shear forces, temperature, pH changes, or other selective factors. However, a rare few protocells may contain a polymer that interacts with the bilayer and thereby stabilizes it against disruption. This is such a strong selective factor that robust early protocells would have been quickly selected according to their content of polymers that protect them against physical stresses.

Another essential function of polymers in the earliest forms of life would be to form pores in the bilayer membrane in order to provide access to potential nutrients available in the surrounding medium. Living cells today have evolved enzymes and channels that transport nutrients and maintain ion concentration gradients, but much simpler peptides can also produce pores. One example is the antibiotic peptide called gramicidin, which has just 15 amino acids and assembles into transient transmembrane channels capable of transporting sodium and potassium ions. This point is highly significant in relation to later adaptation to salty seawater in a marine environment because protocells that survive wet–dry cycles will be preadapted to the osmotic stresses they later experience in seawater. Only protocells that have found ways to withstand osmotic stresses will survive into future generations.

Integrating Polymerization Chemistry with Geological Evidence

Although laboratory studies of chemical reactions can provide clues to the origin of life, it is unlikely that a single breakthrough such as the original Miller experiment or the discovery of the DNA double helix will lead to a eureka moment: "So, *that's* how life began!" Even if an ingenious combination of compounds that assemble into structures having the properties of a living system is discovered in the laboratory, the system will have been created under pristine conditions and would be unlikely to survive and proliferate if exposed to an analogue prebiotic site such as a hot spring or hydrothermal vent.

For this reason, we will propose here a way to integrate laboratory chemistry with geological conditions in natural environments such as the hydration–dehydration cycles occurring in hydrothermal fields associated with volcanism. The narrative was adapted from publications by Damer and Deamer (2015) and Damer (2016).

Plate 10.3 is a local-scale illustration of the geological structures related to volcanism and associated hydrothermal fields. Salty seawater is distilled by evaporation and falls as precipitation on terrestrial volcanic land masses where it forms hydrothermal fields. The inset shows a small boiling pool on Mount Mutnovsky, Kamchatka—an example of a dilute solution that can undergo wet–dry cycles. This small, freshwater pool, which reflects the prebiotic "Goldilocks" chemistry suggested by Powner and Sutherland (2011), supports both polymerization and encapsulation processes that generate protocells. From this pool, a downward flow driven by gravity would bring populations of protocells into new pools with new stresses. This flow ultimately brings increasingly robust populations to dilute rivers and lakes feeding into salty intertidal zones. All of these represent extreme conditions for fragile early living systems that developed in freshwater well supplied with concentrated organics. The downhill journey from fresh to saltwater is entirely opposite to the proposal that life originated in the ocean and only later found its way uphill into freshwater environments on land. In our view, marine hydrothermal vent environments represent a later adaptation for microbial life defined as extremophiles that can thrive in conditions vastly different from the clement Goldilocks pools where life can begin.

The more global-scale illustration in Plate 10.4 shows seven specific steps that were described in earlier chapters and can now be summarized. We will proceed through the steps one by one to describe how they integrate physical and chemical processes presented in earlier chapters with a prebiotic geological landscape.

Synthesis

The setting is the Hadean Earth, approximately 4.2 billion years ago, by which time a global ocean has formed, dotted with emerging volcanic land masses. The solar system remains a dusty place, with the accretion disc still being swept by young planets and large asteroids that experience numerous collisions and impacts. The dust particles are coated with a thin layer of ice containing organic compounds that had been synthesized earlier in their history, along with other compounds adhering to their porous mineral matrix (see Plate 4.5). The organic material was delivered to the Earth's surface as accretionary infall that continues today. Atmospheric photochemistry and volcanic geochemistry also contributed organic compounds to the available inventory. Freshwater hydrothermal pools are optimal conditions that are conducive to the reactions leading to life's origin, while conditions in the marine environment and particularly in hydrothermal vents are extreme. Although we argue that they are implausible sites for life to begin, it is obvious that once begun, microbial life could later adapt to salty seawater and vent conditions

Accumulation

If organic compounds, whether of interstellar or terrestrial origin, fall into the ocean, they will be dispersed into a solution so dilute that they cannot participate in chemical reactions. But if this organic material accumulates on volcanic land masses, precipitation will flush it into a series of smaller collection points and then into hydrothermal pools where it can become sufficiently concentrated to engage in cycles of chemical reactions related to the origin of life (Plate 10.4, left sidebar, a-Organics). The initial collection points are composed of mineral surfaces similar to those seen in hydrothermal fields today. For instance, fresh volcanic basalts resembling a natural version of laboratory glassware would provide containers for aqueous solutions and their organic solutes. Pools will also form when clay particles, or the precipitation of silicate minerals, coat underlying porous surface such as ash and lava. Examples of such pools today are found in hydrothermal fields in Iceland, on Mount Mutnovsky in Kamchatka, in Yellowstone National Park, in Rotorua, New Zealand, and on Mount Lassen in northern California.

Concentration

Organic solutes that are flushed from volcanic mineral surfaces into hydrothermal pools become further concentrated by evaporation, thereby allowing potential reactants to undergo chemical reactions. Assuming that amphiphilic compounds are present in the mixture, the reactants will be organized within multilamellar structures that are ubiquitous in compounds such as fatty acids and their esters. Because hydrothermal pools are chemically complex and vary widely in temperature and pH even within the same field, a product formed in one pool can mix with different compounds formed in a neighboring pool, so chemical processing is not limited to a single pool having one set of conditions. The association of these products with membranous structures such as vesicles sets the stage for combinatorial selection to occur in a cycling environment.

Cycling

Hydrothermal pools undergo wet–dry cycles resulting from evaporation, precipitation, fluctuating water levels, and pulses of hydrothermal fluids from active hot springs and geysers. The large circle in Plate 10.4 summarizes how systems of monomers and their polymers can be coupled through three phases in a periodically cycling pool: (i) a dry phase in which polymers are synthesized within multilamellar structures that form films on mineral surfaces; (ii) a wet phase in which encapsulated systems of polymers are present as protocells which (iii) undergo testing and selection; and (iv) a partially hydrated "gel" phase of crowded surviving protocells that aggregate and interact with one another before drying and fusing into a multilamellar matrix described in the next paragraph.

If amphiphilic compounds are among the solutes in the water along with various monomers, the monomers become concentrated during drying and organized in a liquid crystalline multilamellar matrix. At some point, polymers are synthesized from the monomers, and upon rehydration random sets of polymers are captured in vesicles to form populations of vast numbers of protocells. Frequent cycling of these populations initiates the combinatorial selection process that enables the emergence of increasingly robust systems. As the protocells continue to undergo the stresses of cycling, most will be disrupted and their components dispersed, but a rare few are likely to contain polymers that enhance their survival.

Even if some protocells are successful natural experiments that pass testing during the hydrated phase, that is not the end of the story. Protocells at this stage cannot express catalytic functions nor live independently. For the system to evolve, it must engage in a network in which properties and innovations of molecular selection are shared. At this point the gel phase of the cycle plays a role, because close contacts and fusion events allow sharing of polymers that happen to have essential functions, particularly the ability to create useful metabolic products and catalyze polymerization and replication. At this point the system has reached a form of "prelife" (Plate 10.4, left sidebar, b-Pre-Life). If this seems implausible, it is worth recalling research results from Phil

Holliger's laboratory reporting that ribozymes evolved in the laboratory by selection and amplification were able to catalyze the polymerization of RNA molecules nearly two hundred nucleotides in length (Wochner et al., 2011). Similar results were reported more recently by Horning and Joyce (2016). The products of selection that cycling systems generate can lead to a stepwise emergence of increasingly functional polymers.

The aggregated protocells in the gel phase resemble the progenote concept described by Woese and Fox (1977). Progenotes represent a transitional form between simple chemical systems and the more complex biochemical functions of biology. It is interesting that progenote evolution bears a certain analogy to booting up a computer's operating system in which services start up, share, and support higher level applications. The programs of life are polymer systems selected to perform primitive versions of biological functions: capture of nutrients and chemical energy from the environment; metabolism altering the nutrients into useful compounds; growth by polymerization of monomers; and catalyzed replication of certain polymers. When these systems evolve sufficiently so that they no longer depend on dehydration cycles for polymer synthesis, individual protocells will begin to utilize polymer catalysts to perform these tasks and must therefore pass on coherent tool sets, including primitive genetic instructions to make the tools. An evolutionary chasm will be crossed when protocells begin to divide into daughter cells. Speciation and specialization of cells then becomes possible, crossing the "Darwinian threshold" described by Woese (2002). The tangled roots of the horizontal sharing through the communal progenote now support the growth of multiple interwoven branches as the tree of life continues to evolve. By 3.5 Gya, a common ancestor has emerged with a universal genetic code and a translation system, leaving behind its signature of stromatolites and microfossils in the Pilbara region of Western Australia.

Distribution

Any hypothesis related to the origin of life must propose a pathway for distribution. If the cyclic processes described earlier occurred in an isolated pool in which a living system happened to emerge, there would be no way for it to continue to evolve and adapt to new niches. Hot spring pools on mountainous volcanic landscapes have the obvious advantage that there is a continuous downhill flow from pool to pool, driven by gravity, so that protocell populations can spread to other pools that have entirely distinct characteristics in terms of pH, ionic content, and mineral composition. There they will be subjected to additional stresses, driving them to increasing complexity and finally to cell division and the emergence of primitive life (Plate 10.4, left sidebar, c-Early Life). Dry phase films can also undergo windborne distribution, a mechanism still employed today by desert microbial mat communities. Dehydrated films are an effective preservation and dispersal mechanism for systems of polymers that would be degraded by long-term exposure to hydrolysis, just as seeds and spores preserve DNA between generations of plants and certain microbial populations today. As early cellular populations are physically distributed downhill, they are exposed to a gradient of conditions that force them to become increasingly robust to the kinds of stresses listed earlier in this chapter and ultimately the extreme environment of salty seawater.

Such conditions are observed in hot springs of Yellowstone National Park today. In the upwelling high-temperature spring, thermophilic microbial communities defined as chemotrophic use the energy available in nutrient chemicals, while further down-hill in the cooler and more dilute outflow, photosynthetic organisms capture sunlight as an energy source. Distribution of progenotes and early cellular communities into water with lower concentrations of nutrients would force the invention of some form of photoautotrophy and metabolism as an early adaption. This stepwise adaptation will involve encapsulated polymer systems, but the membranous boundaries can also adapt different compositions that tend to stabilize them. For instance, Black et al. (2013) re-ported that the nucleobase adenine stabilizes fragile fatty acid membranes against salt-induced damage. Groen et al. (2012) reported that polycyclic aromatic hydrocarbons also stabilize fatty acid membranes when incorporated in the bilayers, similar to the effect of another polycyclic compound—cholesterol—on membranes today. An inter-esting possibility is that polycyclic aromatic compounds are components of meteoritic organics and could have provided an early pigment for the capture of energy from sun-light (Deamer, 1992). The Lipid World proposed by Segre et al. (2001) hints at this kind of compositional evolution.

Adaptation

The next stage occurs when autotrophic microbial colonies encounter increasingly brackish water found in estuaries and finally the high salt and divalent cation concen-tration of seawater. Microbial communities today respond to the stress of high salt concentrations by exchanging intracellular sodium chloride for potassium chloride and actively transporting toxic divalent calcium ions out of the cells. This adaptation requires complex, energy-dependent transport of ions by membrane-bound enzymes, suggesting that the first organisms able to live in seawater would have been much more advanced, far from the initial Goldilocks conditions required at the site of life's origins (Mulkidjanian et al., 2012).

Colonization

As microbial communities arrive in the intertidal marine margins, another set of stresses challenges them to adapt. To respond to increased shear forces generated by wave ac-tion, the microorganisms developed cell walls and an extracellular matrix composed of polymers that strengthen adhesion between individual cells in communal aggregates. The aggregates cling to mineral surfaces of tide pools that are washed by tides ten times higher than those of today's oceans. The adhesive polymers that allow the cells to remain fixed on mineral surfaces as microbial mats also binds and cements mineral grains in sediments that settle over them. To have access to sunlight, microorganisms composing microbial mats migrate upward and grow new colonies on top of the cemented sub-structure. Over many years, multiple layers of microbial films and cemented grains produce the laminations characteristic of stromatolites which dominate the earliest fossil record and are still forming today in freshwater and shallow marine environments such as the Hamelin Pool at Shark Bay, Western Australia. At this stage, life would have gained access not only to freshwater environments on land but also to the extensive marine margins of volcanic land masses and protocontinents and would be able to colonize the globe (Plate 10.4, left sidebar, d-Global Life).

The 3.5-billion-year-old fossil stromatolites of the Pilbara represent the final phase of this narrative but also the beginning of life as we know it. The evolutionary process probably required up to a half a billion years to get to stromatolites from the first protocells emerging in hydrothermal fields, and somewhere along that timeline the last universal common ancestors (LUCAs) appeared with a full complement of the functional polymers of life as we know it today. Note the purposeful use of plural in the previous sentence to indicate that LUCA was not a single cell from which all subsequent life descended, but instead would have been in the form of widespread microbial communities that were sharing genetic information. Indeed, applying Morowitz's principal of continuity (1992), the ancestor of the stromatolite is clearly the microbial mat community, but perhaps the microbial community traces its ancestry back to a simpler form of community, the progenote, which in turn got its start as a mass of even simpler protocells.

Summarizing Current Falsification Tests

The results and ideas described in both Chapter 9 and this chapter can be summarized from Popper's perspective. In other words, the scenario described above would tend to be falsified if any of the following tests were unsuccessful:

- Polymers can be synthesized in freshwater by cycles of dehydration and rehydration at elevated temperatures and pH ranges similar to those of acidic hydrothermal fields.
- The synthesis of ester and peptide bonds must be thermodynamically feasible.
- If the monomers are mononucleotides, the polymers will resemble RNA.
- The linking bonds will be phosphodiester linkages that can be hydrolyzed in alkaline pH ranges.
- The polymers accumulate in a kinetic trap.
- The polymers can be purified by methods that are used to purify RNA.
- The polymers can be separated by gel electrophoresis and visualized with dyes that stain RNA
- The polymers will be recognized by enzymes such as alkaline phosphatase and T4 kinase that add a radioactive phosphate from 32-P labeled ATP to the ends of RNA molecules.
- If monomers have the potential to undergo base pairing, such as with AMP and UMP, the polymers will exhibit hyperchromicity.
- The polymers are expected to be single-stranded polyanions, so they should produce ionic current blockades in nanopore analysis.
- If amphiphilic compounds are present in the mixture, the polymers will be encapsulated in the form of protocells.

Each of these tests, if it failed, would in fact falsify the hypothesis. However, all of the expectations survived the experimental tests described in Chapter 9, so the hypothesis has not yet been falsified. But to be convincing, the hypothesis must now undergo more stringent testing such as establishing the molecular structure of the products and demonstrating emergent functions such as catalysis and replication. The predictions described in the next sections will weigh against the hypothesis if they fail experimental testing.

A Plausible Source of Monomers Is Essential

All of the experiments described in Chapter 9 have a tacit assumption that there is a prebiotic source of mononucleotides that are sufficiently stable to stand up to fluctuating hydrothermal conditions at elevated temperatures in acidic media. This naive assumption is highly vulnerable to falsification, but we take comfort from the fact that the assumption seems inescapable. For life to begin, there must have been a process by which mononucleotides or similar monomers were made available for assembly into polymers resembling nucleic acids. The optimistic prediction is that such a process will someday be discovered, most likely as a simpler version of nucleic acids that then evolved over millions of years into the version that functioned in LUCA.

Mononucleotide Mixtures

Life did not begin with RNA composed of a single nucleotide, such as AMP, which is why we use mixtures of AMP and UMP in our experiments. But even this is an over-simplification because we must demonstrate that polymers can be synthesized from a mixture of several different mononucleotides, or at least show how a simpler mixture can evolve into a system capable of initiating and sustaining a genetic code.

Decomposition of Mononucleotides

Although we assume that mononucleotides can survive long enough to undergo polymerization in hydrothermal conditions, we also know that certain reactions occur that decompose mononucleotides. Examples of decomposition include depurination of AMP and GMP and deamination of cytosine to form uracil. Furthermore, a certain amount of caramelization also takes place because a brown polymeric product is also produced in certain conditions. The hypothesis will be falsified if decomposition is so extensive that mononucleotides undergo irreversible damage in hydrothermal conditions such that they are unable to polymerize.

Decomposition of Polymers

Polymers like nucleic acids and proteins also undergo decomposition in living cells and multicellular organisms. For instance, depurination and deamination continuously occur but are repaired by specialized enzymes. Hydrolysis of phosphodiester bonds in DNA is another damaging reaction but again is enzymatically repaired. Proteins also suffer a variety of damaging reactions, both physical (denaturation) and chemical (crosslinking reactions). Microbial populations grow fast enough that most damaging reactions are left behind as undamaged cells proliferate. However, the first forms of cellular life based on RNA and ribozymes probably had much slower growth and division, so it is essential to demonstrate that damage to RNA is not so extensive under hydrothermal conditions that the emergence of life would be impossible. A certain amount of optimism comes from the observation that extremophilic microbial populations can survive temperatures exceeding 100° C. Some can also thrive in acidic and alkaline pH ranges that would be lethal to multicellular organisms.

Plausible Amphiphiles

For experimental convenience, we have used biological amphiphiles such as phospholipids to encapsulate polymers, but these are products of metabolism and obviously not plausible amphiphiles in the prebiotic environment. Therefore, the hypothesis will be falsified if we are unable to demonstrate polymerization in the presence of simpler amphiphiles such as fatty acids or mixtures of fatty acids and their monoglycerides.

Plausible Conditions

In general, it is standard practice for researchers to control experiments by simplifying the conditions, but it is possible to oversimplify and thereby overlook essential variables. For instance, life did not begin under laboratory conditions that use pure organic compounds dissolved in distilled water and maintained in clean glassware at controlled temperature and buffered pH. Instead, as described in the first few chapters, life emerged from complex mixtures of organic solutes undergoing extreme fluctuations of physical conditions including cycles of hydration and dehydration. During our visits to analogue prebiotic environments such as volcanic hydrothermal fields, it became obvious that these are far more complex than any existing laboratory simulation. This is why eighteen factors were listed in Chapter 8 as contributing to the environmental complexity related to the origin of life. Future simulations should attempt to match the complexity of conditions in prebiotic analogue environments. If incremental increases in complexity of protocell systems cannot be demonstrated, the conjecture and related hypotheses described in this chapter will be falsified.

Combinatorial Selection and Evolution of Functional Polymers

If we assume that polymers are synthesized as described in Chapter 9, it is possible that, as well as being encapsulated, some of them will interact with the amphiphilic structures that are present as membranous compartments. We are not specifying the nature of the polymers, but if the solutes include monomers like nucleotides and amino acids, the polymers would resemble RNA and peptides as well as possible complexes of RNA and peptides. It is not difficult to guess what the functional properties of the polymers will be, because they are precursors to functional polymers in cells today. Some general properties are described next, with abbreviations that hint at their functions; the order of the list shows the increasing complexity required for stepwise evolution of protocells toward living systems.

S-polymers have the simplest function, which is to bind to and stabilize a membrane-bounded compartment so that its contents are less likely to disperse into the environment. Examples in cells today are cytoskeletal polymers like spectrin that stabilize erythrocyte membranes.

P-polymers also have one of the simplest functions, which is to form pores in the bilayer membrane that allow access of potential nutrients to the interior volume and equilibration of osmotic pressure gradients. Examples today include short peptides like gramicidin (15mer) and alamethicin (20mer) that form ion-conducting channels in bilayers.

M-polymers catalyze the steps of a primitive metabolism involving chemical reactions among potential nutrients from the external medium after they enter the protocell. The reactions are a source of energy, and the products can be used for polymerization reactions.

R-polymers are able to undergo a primitive version of replication. It may be possible that the monomers not only form polymers under hydrothermal field conditions, but once formed, the polymers can undergo nonenzymatic replication because after the first cycle, any newly synthesized polymer can act as a template. As dehydration occurs, the increasingly concentrated monomers begin to line up on the template by complementary base pairing, and at some point, the reduced water activity drives ester bond synthesis that links them into a second strand.

I-polymers provide a primitive genetic function that first emerges as simple templates for replication catalyzed by R-polymers. However, to support the inheritance of coherent sets of functional polymers and control their actions, more centralized informational polymers must be written, read, and expressed by polymerases that use a coding-translation mechanism.

F-polymers are selected to provide feedback circuits controlling the timing and rates of reactions.

D-polymers initiate and control the division of a protocell following the duplication of its distinct sets of functional polymers and genetic complement.

Each of these polymer classes is in a sense a prediction that can be experimentally tested. The simplest are the S- and P-polymers, but detecting the other functions still lies in the future. Finally, as various functional polymers accumulate in populations of protocells, it becomes possible for the polymers to interact with one another. For instance, interactions between combinations of oligopeptides and oligonucleotides in protocells may produce novel emergent functions that neither polymer has in isolation. Such processes could have given rise to primitive ribosomes.

Conclusion

This chapter was written to show how a conjecture can become a hypothesis following the design of experimental tests. The intention was to provide a guide for our own research, but also to invite other interested researchers to engage in testing the hypothesis, because this is how science progresses. We will conclude the chapter by listing the assumptions that are the foundation of the hypothesis, then summarize a set of predictions that follow.

Assumptions

- The initial reactions can only work in freshwater, not seawater.
- The pH must be in the acidic range.
- Amphiphilic compounds must be present in the mixture along with potential monomers such as mononucleotides and amino acids.
- The mixture must be cycled multiple times between wet and dry conditions.
- The monomers will be concentrated and organized within multilamellar matrices, a condition that promotes polymerization in the dry phase.
- The polymers will be encapsulated upon rehydration.
- The microscopic vesicles and their polymer contents (protocells) can undergo combinatorial selection during wet–dry cycles.

The following predictions and experiments will test the validity of the model:

- It should be possible to demonstrate that distinct populations of encapsulated polymers increasingly tend to persist when lipid vesicles are cycled multiple times between the anhydrous lamellar phase and the hydrated protocell phase.
- At first the monomers will form polymers having random sequences of subunits, but the sequences will become increasingly specific over time as selection occurs.
- When a selective hurdle, such as vesicle survival, is imposed during cycling, polymers will emerge that enhance membrane stability.
- The polymer may stabilize vesicles, but its interaction with lipid could also stabilize the polymer against hydrolysis. This produces a positive feedback loop so that the polymer accumulates during cycling until it dominates the composition of vesicles.
- If the imposed selective hurdle is related to permeability, another polymer species will emerge that allows polar or ionized nutrient solutes to cross the membrane barrier.
- Polymers will begin to exhibit catalytic properties over time, and at some point, a polymerase will emerge that catalyzes replication.
- The polymerase activity could be enhanced by the interaction between two different polymer species such as ribozymes and oligopeptides.
- Populations of functional protocells will become increasingly homogeneous in later cycles as one population with a specific composition is better able to survive selective hurdles or more efficiently uses available resources.

References

Black RA, Blosser MC, Stottrup BL, Tavakley R, Deamer DW, Keller SL (2013) Nucleobases bind to and stabilize aggregates of a prebiotic amphiphile, providing a viable mechanism for the emergence of protocells. *Proc Natl Acad Sci USA* 110, 13272–13276.

Damer B (2016) A field trip to the Archaean in search of Darwin's warm little pond. *Life (Basel)*. doi: 10.3390/life6020021.

Damer B, Deamer D (2015) Coupled phases and combinatorial selection in fluctuating hydrothermal pools: A scenario to guide experimental approaches to the origin of cellular life. *Life (Basel)* 5, 872–887. doi: 10.3390/life5010872.

Deamer DW (1992) Polycyclic aromatic hydrocarbons: Primitive pigment systems in the prebiotic environment. *Adv Space Res* 12, 183–189.

Groen J, Deamer DW, Kros A, Ehrenfreund P (2012) Polycyclic aromatic hydrocarbons as plausible prebiotic membrane components. *Orig Life Evol Biosph* 42, 295–306.

Horning DP, Joyce GF (2016) Amplification of RNA by an RNA polymerase ribozyme. *Proc Natl Acad Sci USA* 113, 9786–9791.

Morowitz H (1992) *Beginnings of Cellular Life: Metabolism Recapitulates Biogenesis*. New Haven, CT: Yale University Press.

Mulkidjanian A, Bychkov A, Dibrova D, Galperin, Koonin E (2012) Origin of first cells at terrestrial, anoxic geothermal fields. *Proc Natl Acad Sci USA*, 109, E821–E830.

Powner MW, Sutherland JD (2011) Prebiotic chemistry: A new modus operandi. *Phil Tran Roy Soc B: Biol Sci* 366, 2870–2877. doi:10.1098/rstb.2011.0134.

Segre D, Deamer DW, Lancet D (2001) The lipid world. *Orig Life Evol Biosph* 31:119–145.

Wochner A, Attwater J, Coulson A, Holliger P (2011) Ribozyme-catalyzed transcription of an active ribozyme. *Science* 332, 209–212.

Woese CR (2002) On the evolution of cells. *Proc Natl Acad Sci USA*, 99, 8742–8747.

Woese CR, Fox GE (1977) The concept of cellular evolution. *J Mol Evol* 10, 1–6.

Where to Next?

Unsolved Questions

Evidence strongly points to ancient ribosomes as self-replicating complexes, where the rRNA in the ribosomes had informational, structural, and catalytic purposes because it could have coded for tRNAs and proteins needed for ribosomal self-replication.

Harry Noller, 2012

Overview

The first ten chapters of this book are a kind of snapshot that captures the current state of knowledge and proposes a scenario for life's beginning that is based on the properties of RNA described by Harry Noller in the epigraph (Noller, 2012). Despite this progress, there are still enormous gaps in our understanding that remain to be filled. The purpose of this chapter is to make those gaps explicit for future investigators who might be attracted to the question of how life can begin. Because life is an interacting system of immense complexity, each component of which is essential to cellular function as a living system, the gaps have little in common. They can be presented as a set of questions related to sources and properties of organic compounds, mechanisms for capturing energy, polymerization and replication of nucleic acids, the origin of ribosomes, and the transmission of genetic information. For each question, I will discuss one or more papers that could provide clues to an answer and then add some ideas that might serve as guides to future research.

What Is a Source of Mononucleotides?

A source of mononucleotides is a problem not just for the hypothesis being presented in this book but for any proposed pathway to the origin of life. Unlike amino acids, there is no obvious source of mononucleotides, for the following reasons: Three different molecular species must be present in an aqueous solution at concentrations sufficient for a reaction to occur. They must somehow, even in this mixture of organic solutes, form a specific ester bond between a phosphate and a ribose, then must form an N-glycoside bond between the ribose and a nucleobase (Fig. 11.1), and these reactions must occur

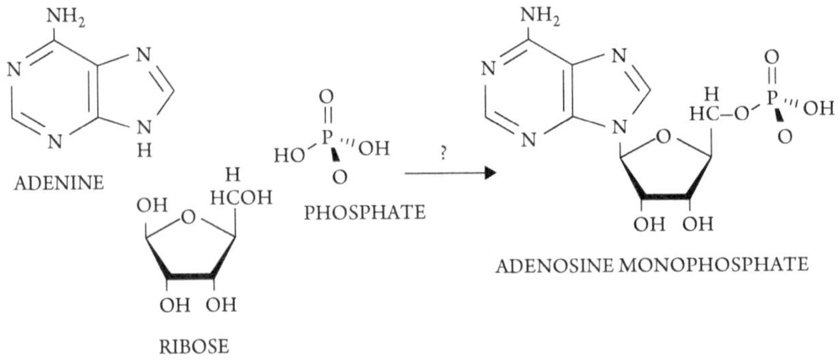

Figure 11.1. Synthesizing a nucleotide monomer of ribonucleic acid is more complicated than synthesizing an amino acid because three different compounds are involved: a nucleobase (like adenine), a ribose, and a phosphate, which must be linked by condensation reactions to form a mononucleotide, in this case adenosine monophosphate.

spontaneously in a hydrothermal environment. Furthermore, it is not enough for one base to occur in the solution, all four (adenine, uracil, guanine, and cytosine) must be present. And finally, the four nucleotides must also be stable enough to undergo polymerization into the RNA-like products described in Chapter 9. A chemist's intuition suggests that this series of reactions is extremely implausible, but if the first forms of life were to exist in an RNA World, something like this needed to happen.

Powner et al. (2009) demonstrated an elegant laboratory synthesis of cytosine monophosphate, but the process depended on multiple steps of organic chemistry that in my judgment are unlikely to occur in the prebiotic conditions of a hydrothermal field. Furthermore, even if a plausible synthesis of mononucleotides is established in the future, decomposition reactions, such as deamination and depurination, will immediately begin, reducing the potential for polymerization into functional nucleic acids. Thus, synthesis of nucleotides will need to proceed at rates that exceed decomposition over long periods of time.

Most earlier work on the origins of life follows the traditional laboratory practice of simplifying conditions to avoid confusion in interpreting the results, so a typical experiment will incorporate just one mononucleotide, such as AMP. However, the genetic code in living cells today requires four different deoxyribonucleotides in DNA (including thymine) and four different ribonucleotides in RNA (in which uracil replaces thymine). Given that the four nucleobases of RNA might accumulate in a hydrothermal pool along with ribose and phosphate and also that hydrolytic decomposition reactions are occurring at a certain rate, how could a polymer like RNA possibly be synthesized? This problem has stymied researchers ever since the RNA World was first proposed. One possible solution—that to my knowledge has not yet been tested—is that the polymers of nucleic acids may represent the lowest energy state in a mixture of nucleobases, with phosphate and ribose undergoing an indefinite series of cyclic condensation and hydrolysis reactions. Stable polymers will, therefore, tend to accumulate in a kinetic trap even in the face of continuing hydrolytic decomposition. The accumulation will not be much affected by the low yields expected in actual prebiotic conditions because cycling conditions allow multiple opportunities for ester bonds to form and persist.

This is one of the most surprising outcomes of the research results described in Chapter 9. There is a general sense that RNA is a fragile polymer, falling apart when exposed to any condition deviating from its normal range of ordinary temperatures and neutral pH. The experimental results presented in this book show that RNA-like polymers not only persist at acidic pH ranges and near-boiling temperatures but can actually be synthesized from monomers under these conditions.

How Did RNA Become Capable of Catalyzed Growth and Replication?

One can imagine that out of a trillion RNA molecules, a few by chance will happen to have polymerase activity. In fact, certain ribozymes evolved under laboratory conditions have this capability (McGinnis and Joyce, 2003; Attwater et al., 2013), but the ribozyme polymerases, like enzyme polymerases, require nucleoside triphosphates (NTPs) as substrates. Before life began, it is unlikely that a source of NTPs was available.

Are there other possibilities? It has been known since Leslie Orgel's early studies in the 1970s that dimers and trimers form when nucleoside monophosphates are simply dried and heated for several days. Much longer strands can be synthesized from mononucleotides undergoing multiple wet–dry cycles in the presence of promoting agents such as phospholipids and monovalent salts, as described in Chapter 9. Could polymerization that produces single-stranded oligonucleotides be extended to a replication process? One possibility is that as soon as single-stranded polymers have been synthesized, they can act as templates. For instance, driven by the wet–dry cycles described in Chapter 9, mononucleotides form small yields of random single-stranded polymers. During the dry phase of later cycles, concentrated mononucleotides may associate with the templates by purine–pyrimidine base pairing. If these undergo polymerization, the result is a duplex molecule in which the sequence of bases in the template will be transferred to the newly synthesized strand as a complementary sequence. The duplex structure unzips in the next wet phase and both strands become available as templates. This would continue in subsequent cycles in a natural version of the polymerase chain reaction. Olasagasti et al. (2011) demonstrated the feasibility of this process with deoxymononucleotides.

Over time, stable polymers with sequences that are more efficient at replicating, perhaps even capable of autocatalysis, will be selected and dominate the mixture. In other words, life did not invent polymerization and replication of nucleic acids. Instead the first protocells incorporated a process that was already occurring and then evolved ribozyme polymerases that made it more efficient.

How Did Biological Homochirality Emerge in a Prebiotic Racemic Mixture?

Many proposals have been advanced to explain how the earliest life became homochiral (see Deamer et al., 2007 for review). Significant progress toward an answer has been made by Donna Blackmond (2010) based on the observation by Soai et al. (1995) that under certain conditions a racemic mixture of compounds can be tipped toward a nearly pure homochiral product. The idea was tested by Hein et al. (2011) who demonstrated that chiral precursors of RNA could in fact be generated by the proposed mechanism.

> The emergence of molecules of single chirality from complex, multi-component mixtures supports the robustness of this synthesis process under potential prebiotic conditions and provides a plausible explanation for the single-handedness of biological molecules before the emergence of self-replicating informational polymers.

Even if homochiral reactants can be synthesized in an enclosed aqueous solution such as a hydrothermal pool, there is a 50/50 chance that they will be either one enantiomer or the other, and they could never be completely pure. It remains to be established how functional polymers can emerge from mixtures of impure chiral reactants. One possibility is that it is simply a numbers game. Imagine a random mixture of racemic monomers that are exposed to the wet–dry cycles described in Chapter 9. If the average length of the polymers is 40 monomers, a single microgram will contain ~30 trillion molecules. Most will have mixed sequences of D and L monomers, but combinatorial arithmetic reveals that a few (~30) will be composed of pure D or pure L monomers. If these fold into structures that resist hydrolysis, or even somehow replicate using monomers of the same chirality as substrates, they will quickly come to dominate the mixture as cycles of synthesis and hydrolysis continue.

Another possibility is that a simple version of life may have been able to use racemic mixtures of monomers, with homochirality evolving later. This was hinted at by Sczepanski and Joyce (2014) who reported a system in which an 83-nucleotide ribozyme composed of all D-nucleotides could catalyze polymerization of L-oligonucleotides on a template of composed of L-nucleotides: "Replicating d- and l-RNA molecules may have emerged together, based on the ability of structured RNAs of one handedness to catalyze the templated polymerization of activated mononucleotides of the opposite handedness."

How Did Life Begin to Use Light as an Energy Source?

The source of virtually all the energy used by life today is visible light captured by chlorophyll. Photosynthesis is the end result of a long chain of evolutionary selection at the molecular level, but there must have been a primitive pigment system that allowed the earliest cellular life to begin to capture light energy. A number of studies have shown that metal sulfides and oxides can transduce light energy into reducing power (Habisreutinger et al., 2013). For instance, Mulkidjanian et al. (2012) suggested that zinc sulfide minerals in hydrothermal fields could have reduced CO_2 to compounds that were incorporated into metabolic pathways.

Chlorophyll is a component of complex photosystems that are embedded in the membranes of chloroplasts and photosynthetic bacteria. Is it possible that primitive

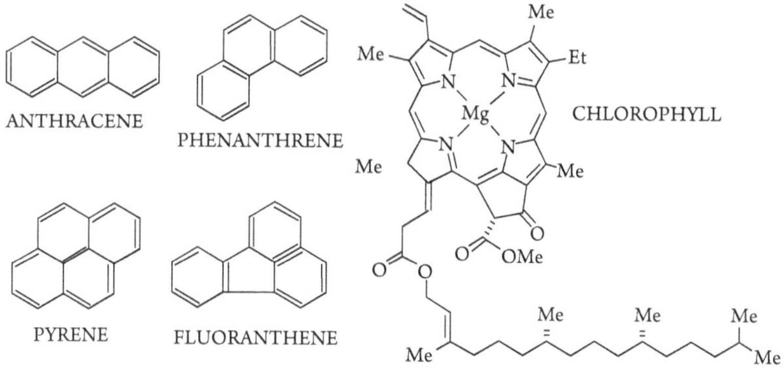

Figure 11.2. Polycyclic aromatic hydrocarbons present in carbonaceous meteorites compared with the structure of chlorophyll, the light trapping pigment of plants today.

photosynthesis used pigments available in the mixture of amphiphilic compounds that assembled into boundary membranes of the first cells? An example of such a pigment today is beta-carotene. Synthesized by plants and an essential vitamin for animal life, beta-carotene becomes retinal after several metabolic steps and is then incorporated into rhodopsin, the primary light-capturing pigment of the retina. But retinal also functions in certain halophilic bacteria as the pigment of bacteriorhodopsin, the simplest protein that can absorb light energy and pump protons across membranes to generate chemiosmotic energy for ATP synthesis (Lanyi, 2004)

One species of organic compounds that could serve as primitive pigment molecules is highlighted here: PAHs, which have been detected as one of the most abundant forms of organic carbon in dense molecular clouds. They are present in carbonaceous meteorites and are likely to be among those compounds delivered to the early Earth during late accretion. Some of the most common PAH derivatives include pyrene, fluoranthene, anthracene, and phenanthrene (Fig. 11.2). Many PAH derivatives absorb near-UV light, then emit the energy as fluorescent wavelengths in the visible range. Significantly, Escabi-Perez et al. (1979) demonstrated that pyrene excited by near-UV photons can deliver electrons to an acceptor in solution—precisely what happens when chlorophyll captures light energy. Furthermore, upon illumination in nonpolar solvents, PAH derivatives such as phenanthrene can bind CO_2 to form carboxylic acids (Tazuke et al., 1986). To my knowledge, this is the simplest photochemical reaction that fixes carbon dioxide. These preliminary results hint at what might be possible, and more extended studies of PAH derivatives as potential photosynthetic pigments are likely to be fruitful.

What Were the First Feedback Loops and How Did They Emerge?

In an important essay, Serhiy Tsokolov (2010) pointed out that all metabolic pathways in life today are controlled by negative feedback loops. Such control mechanisms must have emerged in primitive forms of life, but there has been little interest in developing experimental approaches that would reveal how this might happen. Feedback regulation deserves serious attention in future investigations of prebiotic chemistry related to the origin of life.

What Were the First Steps of Molecular Evolution?

In his book, *The Origin of Species,* Darwin posed a question we are still attempting to answer 150 years later:

> *"Looking at the first dawn of life, when all organic beings, as we may believe, presented the simplest structure, how, it has been asked, could the first steps in the advancement or differentiation of parts have arisen?"*

Jerry Joyce and his students have established a powerful experimental system in which various kinds of catalytic RNA undergo selection and evolution, which is a first step toward answering Darwin's question. For instance, Lincoln and Joyce (2009) wrote:

> Populations of various cross-replicating enzymes were constructed and allowed to compete for a common pool of substrates, during which recombinant replicators arose and grew to dominate the populations. These replicating RNA enzymes can serve as an experimental model of a genetic system.

Horning and Joyce (2016) went on to use such a model genetic system to develop a ribozyme polymerase that can catalyze a 10,000-fold replication and amplification of RNA templates. Results like this bring us a step closer toward demonstrating that an origin of life based on RNA is more than just an idea, and that Darwin's "differentiation of parts" may have involved the emergence and evolution of ribozymes.

How Did Complexes of RNA and Peptides Begin to Function as Ribosomes?

Step by step, a few pioneering researchers are attempting to deduce the pathway by which catalytic ribozymes could evolve into ribosomes. Petrov et al. (2015) established a plausible model of a process by which ribosomes could emerge (Plate 11.1). This paper, with eleven authors from several collaborating research groups, shows how the results from multiple investigators can be integrated into a coherent and testable scenario:

> In this model, the ribosome evolved by accretion, recursively adding expansion segments, iteratively growing, subsuming, and freezing the rRNA.

Functions of expansion segments in the ancestral ribosome are assigned by correspondence with their functions in the extant ribosome. The model explains the evolution of the large ribosomal subunit, the small ribosomal subunit, tRNA, and mRNA. Prokaryotic ribosomes evolved in six phases, sequentially acquiring capabilities for RNA folding, catalysis, subunit association, correlated evolution, decoding, energy-driven translocation, and surface proteinization. . . . The exit tunnel was clearly a central theme of all phases of ribosomal evolution and was continuously extended and rigidified. In the primitive noncoding ribosome, proto-mRNA and the small ribosomal subunit acted as cofactors, positioning the activated ends of tRNAs within the peptidyl transferase center. This association linked the evolution of the large and small ribosomal subunits, proto-mRNA, and tRNA.

But how can we first get to the ancestral RNA shown at the beginning of the pathway in Plate 11.1? Chapter 9 described experimental results showing that wet–dry cycles in the presence of amphiphilic compounds can provide a source of energy for condensation reactions. If mononucleotides are present in the mixture, they polymerize into linear molecules exhibiting chemical and physical properties resembling RNA. Wet–dry cycles at elevated temperatures can also drive synthesis of peptide bonds in mixtures of amino acids to produce oligopeptides (Rodriguez-Garcia et al., 2015). The next experimental question seems obvious: What will happen when mixtures of mononucleotides and amino acids are exposed to wet–dry cycles?

Any knowledgeable chemist will recoil at the complexity of such an experiment, knowing how difficult it will be to analyze products. And yet, young Stanley Miller was bold enough to undertake an experiment of similar complexity. Harold Urey, his supervisor, predicted that he would only achieve "Beilstein," referring to a compendium of all known organic compounds. Urey was correct. Thousands of compounds were synthesized, but the significance of a few major species in the form of amino acids was immediately apparent.

I think that it is time to be similarly bold and begin experimenting with mixtures of mononucleotides and amino acids under conditions in which both monomers are undergoing polymerization. We can start with a single cycle and look for products, but multiple cycles must then be undertaken if the system is to evolve increasing complexity. It seems unlikely that that the products of polymerization will ignore each other. The most interesting outcome would be that the oligonucleotides will begin to interact and form complexes with peptides, the first step toward the assembly of a primitive ribosome. We can only speculate about the properties of such a complex, but it seems possible that a kind of mutual autocatalysis could begin in which the oligonucleotide happened to have a sequence that catalyzed the synthesis of peptide bonds, and the peptides began to catalyze the synthesis of mononucleotides into RNA. Neither RNA nor peptides would be able to act as catalysts by themselves, but the complex would allow them to act on each other.

How Did a Genetic Code Begin to Translate Linear Sequences of Nucleotides in Nucleic Acids into Linear Sequences of Amino Acids in Proteins?

The genetic code is central to all life, and many ideas have been proposed to explain its origin. There is no consensus yet, but our growing understanding of ribosome structure and evolution has led to increasingly detailed proposals about the way that a genetic code can emerge. For instance, Wolf and Koonin (2007) suggested that if ribozymes catalyzed certain reactions of amino acids and peptides, selection for ribozymes would occur that could bind specific amino acids and catalyze the synthesis of peptide bonds. Over time, those ribozyme activities would become the core catalytic function of ribosomes.

> We describe a stepwise model for the origin of the translation system in the ancient RNA world such that each step confers a distinct advantage onto an ensemble of co-evolving genetic elements. Under this scenario, the primary cause for the emergence of translation was the ability of amino acids and peptides to stimulate reactions catalyzed by ribozymes.

More recently, Koonin (2017) expanded on this idea:

> I outline an experimentally testable scenario for the evolution of the code that combines a distinct version of the stereochemical hypothesis, in which amino acids are recognized via unique sites in the tertiary structure of proto-tRNAs, rather than by anticodons, expansion of the code via proto-tRNA duplication, and the frozen accident.

In his paper, Koonin proposed experimental tests that would falsify or confirm the hypothesis. If the tests are successful, they might help answer one of the most important questions related to the origin of life: How did base sequences in nucleic acids begin to incorporate genetic information that could direct the synthesis of proteins?

References

Attwater J, Wochner A, Holliger P (2013) In-ice evolution of RNA polymerase ribozyme activity. *Nat Chem* 5, 1011–1018.

Blackmond D (2010) The origin of biological homochirality. *CSH Persp.* doi: 10.1101/cshperspect. a002147.

Deamer DW, Dick R, Thiemann W, Shinitzky M (2007) Intrinsic asymmetries of amino acid enantiomers and their peptides: A possible role in the origin of biochirality. *Chirality* 19, 751–763.

Escabi-Perez JR, Romero A, Lukac S, Fendler JH (1979) Aspects of artificial photosynthesis: Photoionization and electron transfer in dihexadecyl phosphate vesicles. *J Am Chem Soc* 101, 2231–2233.

Habisreutinger SN, Schmidt-Mende L, Stolarcxyk JK (2013) Photocatalytic reduction of CO_2 on TiO_2 and other semiconductors. *Angew Chem* 52, 7372–7408.

Hein JT, Tse E, Blackmond D (2011) A route to enantiopure RNA precursors from nearly racemic starting materials. *Nat Chem* 3, 704–706.

Horning DP, Joyce GF (2016) Amplification of RNA by an RNA polymerase ribozyme. *Proc Natl Acad Sci USA* 113, 9786–9791.

Koonin EV (2017) Frozen accident pushing 50: Stereochemistry, expansion, and chance in the evolution of the genetic code. *Life (Basel)*. doi: 10.3390/life7020022.

Lanyi JK (2004) Bacteriorhodopsin. *Ann Rev Physiol* 66, 665–688.

Lincoln TA, Joyce GF (2009) Self-sustained replication of an RNA enzyme. *Science* 323, 1229–1232.

McGinnis KE, Joyce GF (2003) In search of an RNA replicase ribozyme. *Chem Biol* 10, 5–14.

Mulkidjanian AY, Bychkov AY, Dibrova DV, Galperin MY, Koonin EV (2012) Origin of first cells at terrestrial, anoxic geothermal fields. *Proc Natl Acad Sci USA* 109. doi: 10.1073/pnas.1117774109.

Noller HF (2012). Evolution of protein synthesis from an RNA world. *Cold Spr Harb Persp in Biol* 4, 1–U20.

Olasagasti F, Kim HJ, Pourmand N, Deamer DW (2011) Non-enzymatic transfer of sequence information under plausible prebiotic conditions. *Biochimie* 93, 556–561.

Petrov AS, Gulen B, Norris AM, Kovacs NA, Bernier CR, Lanier KA, Fox GE, Harvey SC, Wartell RM, Hud NV, Williams LD (2015) History of the ribosome and the origin of translation. *Proc Natl Acad Sci USA* 112, 15396–15401.

Powner MW, Gerland B, Sutherland JD (2009) Synthesis of activated pyrimidine ribonucleotides in prebiotically plausible conditions. *Nature* 459, 239–242. doi: 10.1038/nature08013.

Rodriguez-Garcia M, Surman AJ, Cooper GJT, Suárez-Marina I, Hosni Z, Lee MP, Cronin L (2015) Formation of oligopeptides in high yield under simple programmable conditions. *Nat Comm* 6, 8385.

Sczepanski JT, Joyce GF (2014) A cross-chiral RNA polymerase ribozyme. *Nature* 515, 440.

Soai K, Shibata T, Morioka H, Choji K (1995) Asymmetric autocatalysis and amplification of enantiomeric excess of a chiral molecule. *Nature* 378, 767–768.

Tazuke S, Kazama S, Kitamura N (1986) Reductive photocarboxylation of aromatic hydrocarbons. *J Org Chem* 51, 4548–4553.

Tsokolov SA (2010) Theory of circular organization and negative feedback: Defining life in a cybernetic context. *Astrobiology* 10, 1031–1042.

Wolf YI, Koonin EV (2007) On the origin of the translation system and the genetic code in the RNA world by means of natural selection, exaptation, and subfunctionalization. *Biol Dir*. doi: 10.1186/1745-6150-2-14.

12

Prospects for Life on Other Planets

Life has evolved to thrive in environments that are extreme only by our limited human standards: in the boiling battery acid of Yellowstone hot springs, in the cracks of permanent ice sheets, in the cooling waters of nuclear reactors, miles beneath the Earth's crust, in pure salt crystals, and inside the rocks of the dry valleys of Antarctica.

Jill Tarter

Overview and Questions to Be Addressed

This book describes a hypothetical process in which populations of protocells can spontaneously assemble and begin to grow and proliferate by energy-dependent polymerization. This might seem to be just an academic question pursued by a few dozen researchers as a matter of curiosity, but in the past three decades advances in engineering have reached a point where both NASA and the European Space Agency (ESA) routinely send spacecraft to other planetary objects in our solar system. A major question being pursued is whether life has emerged elsewhere than on Earth. The limited funds available to support such missions require decisions to be made about target priorities that are guided by judgment calls. These in turn depend on plausible scenarios related to the origin of life on habitable planetary surfaces.

We know that other planetary bodies in our solar system have had or do have conditions that would permit microbial life to exist and perhaps even to begin. By a remarkable coincidence, the two most promising objects for extraterrestrial life happen to represent the two alternative scenarios described in this book: An origin of life in conditions of hydrothermal vents or an origin in hydrothermal fields. This final chapter will explore how these alternative views can guide our judgment about where to send future space missions designed as life-detection missions.

Questions to be addressed:

- What is meant by habitability?
- Which planetary bodies are plausible sites for the origin of life?
- How do the hypotheses described in this book relate to those sites?

Habitability: The Goldilocks Principle

There is healthy public interest in how life begins and whether it exists elsewhere in our solar system or on the myriad exoplanets now known to orbit other stars. This has fueled a series of films, television programs, and science fiction novels. Most of these feature extrapolations to intelligent life but a few, such as *The Andromeda Strain*, explore what might happen if a pathogenic organism from space began to spread to the human population. There is a serious and sustained scientific effort—SETI, or Search for Extraterrestrial Intelligence—devoted to finding an answer to this question. Jill Tarter, who directed SETI for many years, noted that even though the conditions on other planets might be what we call extreme, that is no reason to discount the possibility of life, as quoted in this chapter's epigraph. As long as there is liquid water in a temperature range between freezing and boiling and a source of energy, microorganisms have adapted to a variety of extreme niches on our own planet.

So, how extreme can conditions be before they are too extreme? This is the origin of the term Goldilocks Principle: the answer can be as simple as not too hot and not too cold. Both ground-based telescopes and the Kepler satellite telescope have discovered thousands of extrasolar planets, which has led to serious efforts to understand habitability in relation to multiple physical and chemical parameters. The Circumstellar Habitable Zone (CHZ) is based on the assumption that liquid water is essential for any form of life, for which reason the zone is defined in terms of the orbital distance between a planet and its star that would allow liquid water to exist on the planet's surface. The discovery of extrasolar planets around nearby stars—Gliese 581, Trappist-1, and Kepler 90—provided an opportunity to put the CHZ into perspective (Plate 12.1). In our own solar system, for example, the CHZ is the region between the orbits of Venus and Mars, with the Earth and its oceans right in its middle. Both Gliese 581 and Trappist-1 are smaller and cooler than our Sun, so even though their Earth-sized planets' orbits are much closer to those stars, the planets are in a zone where their surface temperatures would allow liquid water to exist. Kepler 90 is slightly larger than the sun with a similar surface temperature, and the number of planets in its orbit—eight—matches the number in our own solar system.

The discovery of liquid water on the moons of Saturn and Jupiter, which are outside the CHZ in our solar system, has forced us to enlarge the definition of habitability beyond simple orbital parameters. In the rest of this chapter we will focus on habitability conditions of icy moons and Mars, then weigh the evidence supporting or falsifying the possibility that life could begin in those conditions.

Enceladus and Europa

The gas giant planets Jupiter and Saturn each have multiple moons. Two spacecraft named Juno and Cassini have been orbiting the planets and visiting their moons for years. (Cassini was caused to dive into Saturn in September 2017 in order to protect the moons from potential contamination by terrestrial microorganisms on the spacecraft.)

The images returned from Enceladus, a small moon of Saturn, and Europa, a larger moon of Jupiter, have been a revelation (Plate 12.2). Both moons are covered by a thick layer of ice, and the ice has lines suggesting a fracturing process related to liquid water beneath.

Cassini provided convincing evidence that there is a subsurface ocean on Enceladus approximately 10 km deep and covered by a 50 km layer of ice. To give a perspective on these numbers, the Earth's ocean has an average depth of just 5 km. What energy source could maintain a liquid ocean on a moon just 500 km in diameter? Lainey et al. (2012) calculated that the friction of tidal heating caused by orbital eccentricity of the moon's orbit around Saturn is sufficient to maintain a liquid ocean.

> Here, we try to determine Saturn's tidal ratio through its current effect on the orbits of the main moons, using astrometric data spanning more than a century. We find an intense tidal dissipation (k2/Q= (2.3 \pm 0.7) \times 10-4), which is about ten times higher than the usual value estimated from theoretical arguments. As a consequence, eccentricity equilibrium for Enceladus can now account for the huge heat emitted from Enceladus' south pole.

What is meant by "huge heat"? The authors are referring to remarkable images returned by Cassini showing plumes, hundreds of kilometers long, being emitted from the south pole (Plate 12.3).

One explanation for the plumes proposes that there is a strong, localized heat source driving hydrothermal activity in the rocky core that is similar to the serpentinization of the Lost City alkaline vents on Earth described in Chapter 3 (McKay et al., 2008; Hsu et al., 2015). In 2008, Cassini managed to fly through the plumes just 50 km above the moon's surface and collect sufficient material for analysis by an onboard mass spectrometer. The plumes contain water, CO_2, methane, and ammonia along with indications of sodium chloride.

A more conservative explanation for the plumes was proposed by Kite and Rudin (2016) who observed diurnal fluctuations in the plume activity. Rather than hydrothermal vents, they proposed that the plumes were caused by tidal forces acting on fractures in the ice through which water could rise and flash evaporate into space.

At this point it does not matter which explanation is closer to the truth. The discovery of organic carbon and ammonia in the plumes has put Enceladus high on the list of sites to be visited by a future life-detection mission.

Hydrothermal Fields on Mars

Of all the rocky planets besides the Earth, Mars has received the most scientific attention, with 44 international missions from 1960 to the present time. The early efforts mostly failed for a variety of reasons, but the Viking mission that launched in 1975 was a major success. The mission included two orbiting spacecraft and two landers that touched down in July and September 1976. In 2003, the Spirit and Opportunity rovers successfully landed on Mars and began to send back the first images of an ancient desert-like landscape. In 2011, the Mars Science Laboratory returned convincing images of an arid, seemingly lifeless planet. Orbiting spacecraft have also revealed numerous volcanoes. One of the oldest is Appolinaris Mons, dated to 3 to 3.5 Gya when Mars still had a sea and a substantial atmosphere. The much-younger Olympus Mons rises two-and-a-half times the height of Mount Everest above the surrounding plane (Plate12.4). It is the largest volcano in the solar system, with eruptions forming the caldera 350 million years ago and lava flows occurring as recently as 2 million years ago.

How Plausible Is It that There Could Be a Separate Origin of Life on Enceladus or Mars?

Jakosky and Shock (1998) were among the first to compare the potential for life to arise on Mars and Europa, using what was known about the early Earth as a guide. Much more has been learned since then, particularly in terms of actual physical and chemical conditions that make a planet habitable. Assuming that there are hydrothermal heat sources driving the plume outbreaks on Enceladus, it is reasonable to think that hydrothermal vents could be present. The facts established by observation are that alkaline vents on Earth have reducing power in the form of dissolved hydrogen with CO_2 as a potential oxidant. As described in Chapter 3, proponents of the vent hypothesis suggest that if CO_2 can be reduced by hydrogen, the products could serve as a feedstock of organic substrates to initiate metabolic processes in primitive life. If we accept the possibility that life could emerge in hydrothermal vent conditions, a life-detection mission to Enceladus, with an instrument package designed to search for biosignatures in the plume composition, can be justified.

My reservations about the plausibility of life beginning under vent conditions apply to both Enceladus and Europa. A salt concentration in the oceans beneath thick ice layers on both moons similar to those in Earth's seawater would inhibit self-assembly of membranous compartments, as described in Chapter 4. Significant thermodynamic hurdles also prevent condensation reactions in aqueous solutions, and an alkaline pH promotes hydrolysis of polymeric linkages such as esters and peptide bonds. Finally, there is no light available under those ice layers for putative microorganisms to develop photosynthesis.

The freshwater alternative described in this book takes into account acidic conditions and hydrothermal fields with wet–dry cycles. These conditions concentrate potential reactants and provide free energy for bond synthesis leading to polymers, with heat providing activation energy. The wet–dry cycles also promote self-assembly of compartments that can encapsulate polymers and form protocells.

Could such conditions have been present on early Mars? The two primary requisite conditions are known to have existed 3–4 billion years ago at the same time that life originated on Earth: shallow seas of liquid water and volcanism that would produce fluctuating pools of freshwater. Furthermore, both the HiRISE (High Resolution Imaging Scientific Experiment) orbiting satellite and the Spirit rover returned images of parts of the planet surface that can reasonably be explained as the remains of hydrothermal fields. One of these was discovered near Columbia Hills by HiRISE and is called Home Plate because its shape resembles the home plate of a baseball diamond

Figure 12.1. Image of Home Plate taken by the HiRISE Mars orbiter. Home Plate is approximately the size of a football field, and evidence obtained by the Spirit rover suggests that it formed in acidic hydrothermal conditions related to volcanic activity. Spirit also discovered opaline silicate minerals nearby, which are known to be produced from silica-bearing water like that of geyser fields in Yellowstone National Park.

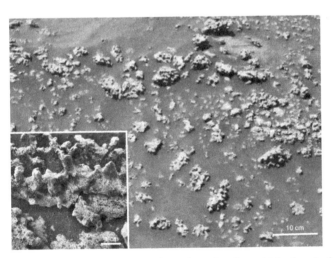

Figure 12.2. Silicate structures protruding from the soil in the neighborhood of Home Plate. Inset shows a magnified view. Image credit Ruff et al. 2011.

(Fig. 12.1). The Spirit rover explored the local neighborhood near Home Plate and returned images showing mineral structures protruding from an otherwise flat surface (Fig. 12.2). These were most likely to have resulted from evaporation of hydrothermal ponds (Ruff et al., 2011), but also bear an intriguing resemblance to stromatolites produced by microbial mats in hydrothermal fields such as those found in Yellowstone National Park.

Conclusions

The submarine hydrothermal vents and volcanic hydrothermal fields that have been proposed as alternative sites for the origin of life on Earth can guide the target priorities and instrument packages of flight missions designed to sample icy plumes of Enceladus, and future landers on Mars. For this reason, we need a better understanding of the chemistry and physics of hydrothermal conditions and how they could promote the origin of life. The following questions can be addressed by laboratory simulations, with experiments designed to test the following essential requirements for life to begin on habitable planetary bodies such as Mars and the icy moons in our solar system:

- A process that sufficiently concentrates diluted reactants.
- Assembly of amphiphilic compounds into membranous compartments.
- A source of chemical energy to drive primitive metabolism and polymerization.
- A mechanism by which polymers can be encapsulated to form protocells.

In the case of icy moons, given the assumption that life can originate in hydrothermal vent conditions on the Earth and that similar conditions exist on Enceladus and Europa, it is feasible that life could be present in their oceans. If so, it might be possible to detect biosignatures in samples of the Enceladus plumes.

On the other hand, the origin of life may require cycles of hydration and dehydration as described in Chapter 9. Because the icy moons lack volcanism and associated land masses, life-detection missions may find that Enceladus and Europa are habitable but sterile.

The same requirements are easily met by the conditions that existed on Mars over 3 billion years ago because fluctuating ponds would be common outcomes of emerging volcanoes and distillation of freshwater from salty seas. The freshwater origin hypothesis presented in this book supports the argument that life-detection missions to Mars should be given priority. The Mars 2020 rover has been designed to investigate this question, and in a few more years we may know whether life exists elsewhere than on Earth.

In *The Origin of Species*, Darwin asked a question: "Looking at the first dawn of life, when all organic beings, as we may believe, presented the simplest structure, how, it has been asked, could the first steps in the advancement or differentiation of parts have arisen?" This book has proposed a possible answer to Darwin's question and extended the answer to the possible origin of life on other habitable planets. Components of this answer fit the definition of a conjecture, but science begins with ideas and conjectures which become hypotheses when they can be tested experimentally or by observation—and finally develop into consensus when testing is successful. There is much testing yet to be done to determine how well the answer proposed here fits the real world.

References

Hsu H-W, Postberg F et al. (2015) Ongoing hydrothermal activities within Enceladus. *Nature* 519: 207–210.

Jakosky BM, Shock EL (1998) The biological potential of Mars, the early Earth, and Europa. *J Geophys Res* 103, 19359–19364

Kite ES, Rubin AM (2016) Sustained eruptions on Enceladus explained by turbulent dissipation in tiger stripes. *Proc Natl Acad Sca USA* 113, 3972–3975.

Lainey V, Karatekin O, Desmars J, Charnoz S, Arlot J-E, Emelyanov N, Le Poncin-Laritte C, Mathis S, Remus F, Tobie G, Zahn J-P (2012) Strong tidal dissipation in Saturn and constraints on Enceladus' thermal state from astrometry. *Astrophys J* 752, 14.

McKay CP, Porco CC, Altheide T, Davis WL, Kral TA (2008) The possible origin and persistence of life on Enceladus and detection of biomarkers in the plume. *Astrobiology* 8, 909–919.

Ruff SW, Farmer JD, Calvin WM, Herkenhoff KE, Johnson JR, Morris RV, Rice MS, Arvidson RE, Bell JF, Christensen PR, Squyres SW (2011) Characteristics, distribution, origin, and significance of opaline silica observed by the Spirit rover in Gusev crater, Mars. *J Geophys Res* 116. doi:10.1029/2010JE003767.

APPENDIX

This book summarizes 40 years of research on the origin of life. The references that follow are for readers who might be interested in the original publications that provided a foundation for the concepts expressed here.

Books

Deamer DW. 2011. *First Life.* Berkeley, CA: University of California Press.

Deamer DW, Fleischaker G (Eds.). 1994. *Origins of Life: The Central Concepts.* Portola Valley, CA: Jones and Bartlett, Publishers.

Deamer DW, Szostak J (Eds.). 2010. *Origins of Life.* Long Island, NY: Cold Spring Harbor Press.

Research Publications

Apel CL, Deamer DW. 2005 The formation of glycerol monodecanoate by a dehydration condensation reaction: Increasing the chemical complexity of amphiphiles on the early Earth. *Orig Life Evol Biosph* 35:323–332.

Black RA, Blosser MC, Stottrup BL, Tavakley R, Deamer DW, Keller SL. 2013. Nucleobases bind to and stabilize aggregates of a prebiotic amphiphile, providing a viable mechanism for the emergence of protocells. *Proc Natl Acad Sci USA* 110:13272–13276.

Chakrabarti A, Deamer DW. 1992. Permeability of lipid bilayers to amino acids and phosphate. *Biochim Biophys Acta* 1111:171–177.

Chakrabarti AC, Deamer DW. 1994. Permeation of membranes by the neutral form of amino acids and peptides. *J Mol Evol.* 39:1–5.

Chakrabarti AC, Joyce GF, Breaker RR, Deamer DW. 1994. RNA synthesis by a liposome-encapsulated polymerase. *J Mol Evol* 39:555–559.

Da Silva L, Maurel MC, Deamer D. 2014. Salt-promoted synthesis of RNA-like molecules in simulated hydrothermal conditions. *J Mol Evol* 80:86–97.

Deamer DW. 1985. Boundary structures are formed by organic compounds of the Murchison carbonaceous chondrite. *Nature* 317:792–794.

Deamer DW. 1992. Polycyclic aromatic hydrocarbons: Primitive pigment systems in the prebiotic environment. *Adv Space Res* 12:183–189.

Deamer DW, Pashley R. 1989. Amphiphilic components of the Murchison carbonaceous chondrite: Surface properties and membrane formation. *Orig Life Evo Bio* 19:21–38.

Deamer DW, Singaram S, Rajamani S, Kompanichenko V, Guggenheim S. 2006. Self-assembly processes in the prebiotic environment. *Philos Trans R Soc Lond B* 361:1809–1818.

De Guzman V, Shenasa H, Vercoutere W, Deamer D. 2014. Generation of oligonucleotides under hydrothermal conditions by non-enzymatic polymerization. *J Mol Evol* 78:251–262.

Dworkin JP, Deamer DW, Sandford SA, Allamandola LJ. 2001. Self-assembling amphiphilic molecules: Synthesis in simulated interstellar/precometary ices. *Proc Natl Acad Sci USA* 98:815–819.

Groen J, Deamer DW, Kros A, Ehrenfreund P. 2012. Polycyclic aromatic hydrocarbons as plausible prebiotic membrane components. *Orig Life Evol Biosph* 42:295–306.

Hargreaves WW, Mulvihill SJ, Deamer DW. 1977. Synthesis of phospholipids and membranes in prebiotic conditions. *Nature* 266:78–80.

Hazen RM, Deamer DW. 2007. Hydrothermal reactions of pyruvic acid: Synthesis, selection, and self-assembly of amphiphilic molecules. *Orig Life Evol Biosph* 37:143–152.

Himbert S, Chapman M, Deamer DW, Rheinstadter MC. 2016. Organization of nucleotides in different environments and the formation of pre-polymers. *Sci Rep* 6:31285.

Kanavarioti A, Monnard P-A, Deamer DW. 2003. Eutectic phases in ice simulate non-enzymatic nucleic acid synthesis. *Astrobiology* 1:271–281.

Lorig-Roach R, Deamer D (2018) Condensation and decomposition of nucleotides in simulated hydrothermal fields. *in Prebiotic Chemistry and Chemical Evolution of the Nucleic Acids.* Cesar Menor-Salvan, ed. Springer Nature Publishing.

Maurer SE, Deamer DW, Boncella JM, Monnard P-A. 2009. Chemical evolution of amphiphiles: Glycerol monoacyl derivatives stabilize plausible prebiotic membranes. *Astrobiology* 9:979–987.

Mautner MN, Leonard R, Deamer DW. 1995. Meteorite organics in planetary environments: Hydrothermal release, surface activity and microbial utilization. *Planet Space Sci* 43:139–147.

Milshteyn D, Damer B, Havig J, Deamer D (2018) Amphiphilic compounds assemble into membranous vesicles in hydrothermal hot spring water but not in seawater. *Life (Basel)* doi: 10.3390/ life8020011

Monnard P-A, Apel CL, Kanavarioti A, Deamer DW. 2002. Influence of ionic inorganic solutes on self-assembly and polymerization processes related to early forms of life: Implications for a prebiotic aqueous medium. *Astrobiology* 2:139–152.

Monnard P-A, Deamer DW. 2002. Membrane self-assembly processes: Steps toward the first cellular life. *Anat Rec* 268:196–207.

Monnard P-A, Deamer DW. 2001. Nutrient uptake by protocells: A liposome model system. *Orig Life Evol Biosph.* 31:147–155.

Monnard P-A, Deamer DW. 2003. Preparation of vesicles from non-phospholipid amphiphiles. *Methods Enzymol* 372:133–151.

Monnard P-A, Kanavarioti A, Deamer DW. 2003 Eutectic phase polymerization of activated ribonucleotide mixtures yields quasi-equimolar incorporation of purine and pyrimidine nucleobases. *J Am Chem Soc* 125:13734–13740.

Monnard P-A, Luptak A, Deamer DW. 2007. Models of primitive cellular life: Polymerases and templates in liposomes. *Philos Trans R Soc Lond B* 362:1741–1750.

Namani T, Deamer DW. 2008. Stability of model membranes in extreme environments. *Orig Life Evol Biosph* 38:329–341.

Oliver A, Deamer DW. 1994. Alpha helical hydrophobic polypeptides form proton-selective channels in lipid bilayers. *Biophys J* 66:1364–1379.

Paula S, Volkov AG, Van Hoek AN, Haines TH, Deamer DW. 1996. Permeation of protons, potassium ions, and small polar molecules through phospholipid bilayers as a function of membrane thickness. *Biophys J* 70:339–348.

Rajamani S, Vlassov A, Benner S, Coombs A, Olasagasti F, Deamer DW. 2008. Lipid-assisted synthesis of RNA-like polymers from mononucleotides. *Orig Life Evol Biosph* 38:57–74.

Simoneit B, Rushdi AI, Deamer DW. 2007. Abiotic formation of acylglycerols under simulated hydrothermal conditions and self-assembly properties of such lipid products. *Adv Space Res* 11:1649–1656.

Toppozini L, Dies H, Deamer DW, Rheinstädter MC. 2013. Adenosine monophosphate forms ordered arrays in multilamellar lipid matrices: Insights into assembly of nucleic acid for primitive life. *PLoS ONE* 8: e62810.

Zepik HH, Rajamani S, Maurel MC, Deamer DW. 2007. Oligomerization of thioglutamic acid: Encapsulated reactions and lipid catalysis. *Orig Life Evol Biosph* 37:495–505.

Reviews

Damer B, Deamer DW. 2015. Coupled phases and combinatorial selection in fluctuating hydrothermal pools: A scenario to guide experimental approaches to the origin of cellular life. *Life (Basel)* 5:872–887. doi: 10.3390/life5010872.

Deamer DW. 1986. Role of amphiphilic compounds in the evolution of membrane structure on the prebiotic Earth. *Orig Life Evol Biosph* 17:3–25.

Deamer DW. 1997. The first living systems: A bioenergetic perspective. *Microbiol Mol Biol Rev* 61:239–262.

Deamer DW. 2003. Self-assembly and energy flow through molecular systems: The origin of cellular life. In W Sullivan (Ed.), *Astrobiology*. London: Cambridge University Press.

Deamer DW. 2009a. First life and next life. *MIT Technology Review,* May/June.

Deamer DW. 2009b. On the origin of systems. Systems biology, synthetic biology and the origin of life. *EMBO Rep. 10 Suppl 1*:S1–S4.

Deamer DW. 2012. Liquid crystalline nanostructures: Organizing matrices for non-enzymatic nucleic acid polymerization. *Chem Soc Rev* 41:5375–5379.

Deamer DW. 2017. Conjecture and hypothesis: The importance of reality checks. *Beilstein J Org Chem* 13:620–624.

Deamer DW, Damer D. 2017. Can life begin on Enceladus? A perspective from hydrothermal chemistry. *Astrobiology* 17:834–839.

Deamer DW, Dick R, Thiemann W, Shinitzky M. 2007 Intrinsic asymmetries of amino acid enantiomers and their peptides: A possible role in the origin of biochirality. *Chirality* 19:751–763.

Deamer DW, Dworkin J. 2005. Chemistry and physics of primitive membranes. In P Walde (Ed.), *Prebiotic Chemistry: From Simple Aamphiphiles to Protocell Models* (pp. 1–27). Berlin: Springer: 2005.

Deamer DW, Dworkin JP, Sandford SA, Bernstein MP, Allamandola LJ. 2002. The first cell membranes. *Astrobiology.* 2:371–382.

Deamer DW, Georgio CD. 2015. Hydrothermal conditions and the origin of cellular life. *Astrobiology* 15:1091–1095. doi:10.1089/ast.2015.1338.

Deamer DW, Harang-Mahon E, Bosco G. 1994. Self-assembly and function of primitive membrane structures. In S Bengston (Ed.), *Early Life on Earth. Nobel Symposium No. 84*. New York: Columbia University Press.

Deamer DW, Oro J. 1980. Role of lipids in prebiotic structures. *Biosystems* 12:167–175.

Deamer DW, Weber AL. 2010. Bioenergetics and life's origins. *Cold Spring Harb Perspect Biol* 2:a004929.

Georgiou CD, Deamer DW. 2014. Lipids as universal biomarkers of extraterrestrial life. *Astrobiology* 14:541–549. doi: 10.1089/ast.2013.1134.

Hargreaves WR, Deamer DW. 1978. Origin and early evolution of bilayer membranes. In DW Deamer (Ed.), *Light Transducing Membranes: Structure, Function and Evolution* (pp. 23–59). New York: Academic Press.

Himbert S, Chapman M, Deamer DW, Rheinstadter MC. 2016. Organization of nucleotides in different environments and the formation of pre-polymers. *Sci Reps* 6. doi:10.1038/srep31285.

Monnard P-A, Deamer DW. 2002. Membrane self-assembly processes: Steps toward the first cellular life. *Anatoml Rec* 268:196–207.

Morowitz HJ, Deamer DW, Smith T. 1991. Biogenesis as an evolutionary process. *J Mol Evol* 33:207–208.

Morowitz HJ, Heinz B, Deamer DW. 1988. The chemical logic of a minimum protocell. *Orig Life Evol Biosph* 18:281–287.

Pohorille A, Deamer DW. 2009. Self-assembly and function of primitive cell membranes. *Res Microbiol* 160:449–456.

Ross D, Deamer DW. 2016. Dry/wet cycling and the thermodynamics and kinetics of prebiotic polymer synthesis. *Life (Basel)*. doi: 10.3390/life6030028.

Segre D, Deamer DW, Lancet D. 2001. The lipid world. *Orig Life Evol Biosphere* 31:119–145.

Van Kranendonk M, Deamer D, Djokic T. 2017. Life springs. *Scientific American* August.

INDEX